Also by Harvey T. Dearden:

*Functional Safety in Practice*
*Crowns & Coronets, Mitres & Manes*

I THINK THE COMPUTER GOT THE DECIMAL POINT IN THE WRONG PLACE FOR THE NEW MARK SIX!

# PROFESSIONAL ENGINEERING PRACTICE:

## REFLECTIONS ON THE ROLE OF THE PROFESSIONAL ENGINEER

Harvey T. Dearden CEng

*J. R. Smith* PUBLISHING

J. R. Smith Publishing

First published in paperback in Great Britain in 2013 by Harriet Parkinson Publishing, Manchester.

Second edition published in 2017 by J. R. Smith Publishing, London.

Copyright © 2017 Harvey T. Dearden

Harvey T. Dearden asserts his moral right to be identified as the author of this work.

All rights reserved.

No part of this publication may be reproduced, stored in a retrieval system, or transmitted, in any form or by any means, electronic, mechanical, photocopying, recording or otherwise, without the prior permission of the author.

ISBN-13: 978-1542730747
ISBN-10: 1542730740

Printed and bound by CreateSpace, a DBA of On-Demand Publishing, LLC

In memory of my father,
who taught me much about engineering
and even more about life.

# CONTENTS

### (CHAPTER 16 ONWARDS NEW TO THIS EDITION)

| | | |
|---|---|---|
| 1 | Engineering Judgement | 1 |
| 2 | Risk Assessment Tools: Good Servants, Poor Masters | 7 |
| 3 | And Yet it Moves! | 12 |
| 4 | Professionalism and the Engineer | 15 |
| 5 | Bring Back the Slide Rule? | 19 |
| 6 | Grey is Good! | 22 |
| 7 | An Inspector Calls... | 26 |
| 8 | Codes of Conduct: Strait Jackets or Shields? | 30 |
| 9 | Touching Wood: Hope and the Professional Engineer | 33 |
| 10 | The Measure of an Engineer? | 36 |
| 11 | Safety Management: Beware the Reality Gap | 41 |
| 12 | Engineering, Fast and Slow | 44 |
| 13 | Risk Assessment and Fermi Calculations | 50 |
| 14 | Staying Aligned | 54 |
| 15 | Jane Austen: Engineer? | 59 |
| 16 | Beyond ALARP | 62 |
| 17 | Internet Forums for Professional Groups: Uses and Abuses | 64 |

| 18 | Logical V Sensible | 67 |
| 19 | Maths for Maths' Sake | 69 |
| 20 | Professional Competence | 71 |
| 21 | Crossing the Line? | 75 |
| 22 | To Shrug or Not to Shrug? | 78 |
| 23 | The Dumb Customer | 80 |
| 24 | Engineering Judgement Space: Four Shades of Grey? | 83 |
| 25 | Engineering a Joke? | 88 |
|    | Coda: To Boldly Engineer | 90 |
|    | Appendix: Twenty Things to Remember | 92 |

# PREFACE TO THE SECOND EDITION

I trust both developing and experienced engineers will find this expanded collection useful. The individual essays are unlikely to be found to outstay their welcome; whatever else may be said of them, they are short and to the point.

I am all admiration for the accomplished novelist; his/her ability to expansively provide setting, characterisation, dialogue and 'colour' eludes me - my instinct is for concision, to pare the superfluous and reveal the pertinent essence.

That said, I would argue that this is a fitting instinct for the professional engineer. We should seek concision in our thinking and expression – and in the thinking and expression of those we work with. In this way, we can improve the 'signal-to-noise' ratio in our work.

We might invoke a variant of Occam's razor:

*Frustra fit per plura quod potest fieri per pauciora*

(It is futile to do with more things, that which can be done with fewer.)

Not a useful maxim for the novelist, but for the professional engineer...

Harvey T. Dearden CEng
North Wales 2017

# PREFACE TO THE FIRST EDITION

It seems that engineering students are not taught about the nature of the engineering profession and the role of engineering judgement; the expectation appears to be that they will acquire the necessary understanding through experience and extended exposure to good practice. I would agree that a full understanding of these concerns can only be acquired through experience, but I also believe that appropriate teaching will catalyse learning from experience by helping the developing engineer to distinguish good practice from bad. Once this facility is established, even exposure to bad practice can be good (useful) experience; without it, such exposure will be damaging. The damage may, if unrecognised, prove irreversible; at best, it will retard development. These essays explore these concerns and will hopefully provide a useful foundation on which to build through experience. For the mature engineer, it is hoped that they will provide a useful distillation of learning. My own experience has been predominantly within the process industry sector and many of the specific examples I point to are drawn from that sector, nevertheless, the general points should find application in any engineering discipline.

This book has been compiled from a number of articles, some of which may be familiar, having appeared in edited form in engineering institution magazines. Note that throughout I employ the convention that the male pronouns may be assumed to apply equally to the female of the species.

Harvey T. Dearden CEng
North Wales 2013

# A NOTE CONCERNING THE UK ENGINEERING PROFESSION

Some of the following essays refer to circumstances prevailing in the UK. Readers from outside the UK may be surprised to learn that the title 'Engineer' is not protected in the United Kingdom (once regarded as the workshop of the world!). Anyone, regardless of qualifications, may style himself as an 'engineer'. For the man in the UK street, anyone that wields either a spanner or a screwdriver is an 'engineer'.

There is a standing joke about two eligible bachelors, both professional men (one a medical doctor and one an engineer), going to a dinner party. On meeting the doctor and learning of his profession, the host introduces him to his daughter. On meeting the engineer and learning of his profession, the host introduces him to his washing machine!

That said, the titles Chartered Engineer (CEng) and Incorporated Engineer (IEng), as administered by the Engineering Council UK, are protected and only awarded on demonstration of appropriate qualifications and competence.

Many of these essays discuss issues of safety, risk assessment and management. The HSE (Health and Safety Executive) is the national independent regulator for matters of work-related health, safety and illness. Environmental concerns are regulated by the EA (Environment Agency) and the SEPA (Scottish Environment Protection Agency).

# 1
# ENGINEERING JUDGEMENT

What is the implication of using engineering judgement? Is it that rigour is sacrificed on the altar of expediency? Sometimes there is a trade-off to be made in this respect, but there is something of a false dichotomy here. Often it is a question of pragmatism rather than expediency. Rigour is all very well, but not always a practicable option and the law of diminishing returns will once more hold sway. The truth is that we always employ engineering judgement in our evaluations; it is simply a matter of degree. We know that 3 is bigger than 2 with absolute mathematical rigour, but if we have a structure A that we evaluate has a load capability of 3 and a structure B that we evaluate as having a capability of 2, we cannot, with absolute rigour, state that structure A is actually stronger than structure B. Whenever we consider different courses of action, different design proposals or consider equipment/material options, there will be some uncertainty.

We could say that design A is stronger than design B 'all else being equal', but there would remain some possibility of some unrevealed defect/flaw or unrecognised circumstance that could compromise the actual strength of structure A. We may be supremely confident that A will be stronger than B, but we cannot know this absolutely until we actually load the structures. It might be argued that absolute rigour is achieved when our evaluations are traceable to international standards of measurement and fundamental natural laws, but even then judgement will be invoked in deciding which laws and measurements are relevant and what the acceptable levels of uncertainty are.

Any attempt to pursue absolute rigour and avoid engineering judgement is doomed to failure. There will always be a degree of uncertainty about our assessment/design/approach or about the appropriate criteria of adequacy. We may have recourse to published standards and codes of practice, but these will themselves contain judgements. The Engineering Council's publication, *Guidance on Risk for the Engineering Profession*[1], says:

> ...*Regulations and codes are generic. They can only deal with anticipated events, and cannot predict every possible situation. Engineers should take a measured, yet challenging approach to potential risks, whether or not regulations apply (...) Engineers should recognise and understand the intent behind standards and codes, and understand when their limits are being approached...*

Our concern is not to avoid judgement but rather to exercise it appropriately, underpinned with sufficient rigour to establish a suitable level of confidence. Absolute confidence is something we can only approach asymptotically however, and therefore the question arises: what constitutes 'a suitable level of confidence'? This itself requires an exercise in judgement and will depend on the potential consequences of failure. We are faced with an infinite regression here.

There is, of course, an obligation on professional engineers not to exceed their level of competence and to seek assistance where necessary. But avoidance of the exercise of judgement when within our competence is an abdication of professional responsibility. Unwarranted conservatism is often pursued in an attempt to eliminate any question of personal liability or possible criticism, but this will spuriously divert resources and undermine professional credibility when there is seen to be a disconnect between provisions and requirements. If we believe our judgement to be soundly based, we should stand ready to defend it. The approach to 'absolute confidence' may be asymptotic but it is stepwise, not continuous. There comes a break point where the marginal increment in confidence will come at grossly disproportionate expenditure of resources. What is particularly

---

[1] *Guidance on Risk for the Engineering Profession* (London: Engineering Council, March 2011).

disappointing is the practice of 'going through the motions' when there is understood to be no benefit or even when it may be counter-productive. How then may we acquire a facility for engineering judgement? From what may it be drawn?

It derives from established precedent and an understanding of the mechanisms (in the widest sense of causal relationships) that are relevant in a given context, together with an awareness of the sensitivity of our evaluations to uncertainty in raw data or inherent assumptions. Precedents are often enshrined in standards and codes of practice, but we need an understanding of these precedents and standards as applied in a variety of contexts in order to develop a sense of fitness for purpose. We may extrapolate/interpolate from established precedent but we need an understanding of the relevant mechanisms. The builders of mediaeval cathedrals were fortunate in the very high compressive strength of stone; they were able to extrapolate to ever-bigger structures with impunity - up to a point! They had no appreciation that as they scaled up their structures the loads increased as the cube of scale and the cross sections as the square, and therefore that the stresses increased proportionately. We might interpolate between vibration levels at two speeds but if we have no understanding of resonance and critical speeds, we may court disaster. We may substantiate our evaluations with calculations but we will exercise judgement about which calculations, on what basis and what level of precision. There can be no absolute confidence.

Ultimately, for the individual exercising judgement, the level of confidence will come down to some psychological measure derived from a sense of 'fitness for purpose'. We form an impression of adequacy, but how may the 'calibration' of this psychological variable be ascertained? A fool may be supremely confident! It becomes a societal question of professionalism - we rely on the endorsement of our professional peers. The further removed from rigour we find ourselves, the more we should invite the view of our peers. The value of a 'sanity/reality check' is not to be underestimated.

It seems there is an increasing reliance placed on certification of qualifications, but our confident fool may well be armed with a certificate! Certification of the personal qualities that manifest in effective judgement is not available - and experience does not necessarily translate into wisdom. (My own experience is that beyond a certain point there may well be a negative correlation between qualifications and wisdom.) It is through a range of experience, both direct and indirect, that we establish intuitive benchmarks against which we may evaluate proposals. It is by examining engineering history that we acquire indirect but nonetheless critical experience that helps to build capability in engineering judgement. Historical failures are particularly instructive. Beware the engineer that only has time for the state of the art and no interest of the roots of his discipline.

The Space Shuttle Challenger disaster is an object lesson. The root cause was found to be an O-ring seal that failed in service because it did not retain its resilience in the cold conditions prevailing at the time of the launch. The Rogers Commission enquiry[2] into the disaster revealed that engineers were pressured to 'put their business hats on' when making recommendations about whether to proceed with the launch; effectively they were urged to make a business judgement rather than an engineering one. If only they had kept their 'engineering hats' on. Any professional engineer engaged in a commercial enterprise is essentially by definition called upon to wear both hats at the same time. The commission concluded:

> ...*failures in communication (...) resulted in a decision to launch 51-L based on incomplete and sometimes misleading information, a conflict between engineering data and management judgments, and a* NASA *management structure that permitted internal flight safety problems to bypass key Shuttle managers...*

Interesting that Richard Feynman (a Nobel laureate and professor of theoretical physics) was appointed to the commission - clearly the disaster was not the domain of theoretical physics. Feynman however, did possess an enquiring mind, an understanding of the relevant mechanisms and a maturity of outlook. His independent frame of mind placed him at odds with others on the commission; his particular views concerning the safety culture within NASA were added to the report as Appendix F, *Personal observations on the reliability of the Shuttle.* I include here an extract from Feynman's appendix[*] that strikes me as

---

[2] *Report to the President by the Presidential commission on the Space Shuttle Challenger Accident* (June 6th 1986).

[*] For the full story of Feynman's involvement, including a copy of the appendix, see *What Do You Care What Other People Think*, Richard P. Feynman (London: Penguin Books, 2007).

particularly telling:

> *The origin and consequences of the erosion and blow-by were not understood. They did not occur equally on all flights and all joints; sometimes more, and sometimes less. Why not sometime, when whatever conditions determined it were right, still more leading to catastrophe?*
>
> *In spite of these variations from case to case, officials behaved as if they understood it, giving apparently logical arguments to each other often depending on the "success" of previous flights. For example, in determining if flight 51-L was safe to fly in the face of ring erosion in flight 51-C, it was noted that the erosion depth was only one-third of the radius. It had been noted in an experiment cutting the ring that cutting it as deep as one radius was necessary before the ring failed. Instead of being very concerned that variations of poorly understood conditions might reasonably create a deeper erosion this time, it was asserted, there was "a safety factor of three". This is a strange use of the engineer's term, "safety factor". If a bridge is built to withstand a certain load without the beams permanently deforming, cracking, or breaking, it may be designed for the materials used to actually stand up under three times the load. This "safety factor" is to allow for uncertain excesses of load, or unknown extra loads, or weaknesses in the material that might have unexpected flaws, etc. If now the expected load comes on to the new bridge and a crack appears in a beam, this is a failure of the design. There was no safety factor at all; even though the bridge did not actually collapse because the crack went only one-third of the way through the beam. The O-rings of the Solid Rocket Boosters were not designed to erode. Erosion was a clue that something was wrong. Erosion was not something from which safety can be inferred.*
>
> *There was no way, without full understanding, that one could have confidence that conditions the next time might not produce erosion three times more severe than the time before. Nevertheless, officials fooled themselves into thinking they had such understanding and confidence, in spite of the peculiar variations from case to case. A mathematical model was made to calculate erosion. This was a model based not on physical understanding but on empirical curve fitting.*

Hindsight is often facetiously said to be 20/20 but it isn't true and 'those that fail to heed the lessons of history are doomed to repeat it'. I do not believe there was any conspiracy or wilful negligence in the Challenger story; there were failures in communication and understanding. There was a rationale behind the evaluations that were made; the individual elements made sense, the arguments seemed plausible - the failure was in logic. As Feynman points out, 'officials fooled themselves into thinking they had such understanding and confidence'. Judgement was being exercised, but without a proper understanding of the relevant mechanisms.

In order to develop a facility for engineering judgement, a questioning mind is needed, coupled with a maturity that does not always come with years. Those that see engineering as a job (merely a question of going through the prescribed motions) rather than a profession are unlikely to exhibit the appropriate characteristics.

# 2
# RISK ASSESSMENT TOOLS: GOOD SERVANTS, POOR MASTERS

Risk assessment tools should come with a health warning: 'uncritical use may seriously damage your business'. Worst-case assumptions may appear prudent, but their undiscriminating use may seriously undermine the value of a risk assessment in identifying the appropriate allocation of resources. There are a variety of approaches that may be adopted in undertaking risk assessments in support of the development of a safety case or SIL (Safety Integrity Level) determination. There is a spectrum of possibilities ranging from the purely qualitative, through semi-quantitative, to the fully quantified. In assessing risk, the HSE (Health and Safety Executive in the UK) employ the concept of 'proportionality'; risk assessments and adopted risk reduction measures are required to be 'proportionate' to the risk. In essence, the greater the risk, the greater the required degree of rigour in the risk assessment and the demonstration that risks are ALARP (As Low As Reasonably Practicable), and the more a duty holder would be expected to pay to reduce those risks.

A risk may be characterised as having 'high proportionality' if it is assessed as approaching the intolerable region of the 'risk triangle', which categorises risk as being 'broadly acceptable', 'tolerable if ALARP' and 'intolerable'.

Proportionality should not be confused with 'proportion factor', which is the ratio of the CPF (Cost of Preventing a Fatality) and the

VPF (Value of Preventing a Fatality) and which may be used in assessing whether any further expenditure on incremental risk reduction would be 'grossly disproportionate' as a test of whether a risk was ALARP.

Some of the available risk assessment techniques are highly refined but this should not blind us to the inevitable uncertainties in the results. A rigorous process will not compensate for uncertain data and any quest for absolute accuracy is doomed. A particular problem is the compounding of conservatism: a succession of 'worst-case' assumptions that are thought to represent an upper bound to risk will quickly generate gross distortion of the overall assessment. It may well seem prudent to 'err on the side of caution' but do this at successive points and your analysis may well be one or more orders of magnitude from a 'true' estimate. It might be argued that the distortion is 'safe' but if it causes a misdirection of resources or an unwarranted elaboration of provisions which are more difficult to manage and maintain, the result may well be an overall net loss of safety in practice.

A wrongheaded and naïve approach is to perform a risk assessment and unquestioningly accept the outcome as definitive; the outcome should always be critically reviewed to see if it appears sensible. It may be that the assessment reveals hitherto unrecognised levels of risk or identifies an element as being of more significance than was previously understood. It is perhaps when revealing any such anomalies that these tools are at their most useful. If the outcome does not fit with your experience or seems to suggest that what you understood to be established good practice is inadequate, you should probe further to discover whether it is your expectations or the assessment that is flawed. If your SIL determination process reveals wholesale design deficiencies, your first thought should be to check the calibration of your approach.

The tools at the less rigorous end of the spectrum e.g. risk graphs and matrices tend to be relatively coarse in their resolution of risk and are more susceptible to inadvertent compounding of conservatism. These less rigorous tools might be effectively deployed as a first pass filter with apparently higher risk scenarios 'parked' for a higher resolution and higher rigour assessment. The key point is that risk assessment should be performed to validate your judgement (which should itself be suitably informed by training, experience, knowledge of good practice and an understanding of pertinent mechanisms) rather than to attempt to directly and definitively establish the 'true' level of risk.

It is in recognition of this that the Engineering Council's publication, *Guidance on Risk for the Engineering Profession*[1], says "...bear in mind that risk assessment should be used as an aid to professional judgement and not as a substitute for it". It may be that after critical probing to find the reason for any discrepancy between your expectation and the assessment, you may be obliged to 'recalibrate' your judgement. Embrace the opportunity; this is quality CPD (Continuing Professional Development).

The 'new' risks must then be addressed in an appropriate, responsible manner. Clearly what is not acceptable is to cynically manipulate the risk assessment to remove any embarrassment or inconvenience. On the other hand, your critical review may find that the assessment is flawed in some way. Perhaps it is through inadvertent compounding of conservatism or some critical omission; perhaps a missing conditional modifier or enabling event; or some circumstance that should be factored in to the analysis to qualify the true hazard potential. For example, a flammable release risk should be qualified based on the likely size of the release, the presence and vulnerability of people, the likelihood of a source of ignition, the likelihood of detection and effective intervention, the likelihood of the absence of a good stiff breeze etc.

Very often it will be self-evident that the probability of a given circumstance will be less than 100% but there may be no ready means of evaluating how much less. The default position is therefore to conservatively assume a probability of 100%. This is not unreasonable in itself, but if repeated a number of times we may suffer gross distortion in our final estimate. It may be that in recognition of this 'default' conservatism we might choose <u>not</u> to be conservative in respect of some other provision, which is more tractable in estimation.

A good example is the guidance given in the Buncefield report[2] in terms of the 'probability of explosion after ignition' as a conditional modifier to be used in LOPA (Layer of Protection Analysis) studies:

---

[1] *Guidance on Risk for the Engineering Profession* (London: Engineering Council, March 2011).

[2] *"Safety and environmental standards for fuel storage sites" Final Report* (Surrey: HSE, 2009).

*...Given the present state of knowledge about the Buncefield explosion mechanism this report tentatively proposes that the value of this modifier should be taken as unity in the stable, low wind-speed, conditions that are the basis of this hazardous scenario. A much lower, and possibly zero, probability might be appropriate. It is possible that an improved understanding of the explosion mechanism may allow a better basis for determining the value of this factor in the future...* (Appendix 2, clause 141)

The explosive (as distinct from flash fire) nature of the Buncefield incident was something of a surprise. The implication would seem to be that the circumstances were unlikely, otherwise there would have been no surprise; the phenomenon would have been well known and the possibility recognised. If the actual probability does indeed approach zero, the assessment of the risk associated with an explosion will have been inflated by a factor 'approaching' infinity!

This highlights the need to be wary of accepting risk assessment quantification as providing any measure of absolute risk; these tools are typically more useful in providing a measure of relative risk and providing a means of comparison. This is of particular significance if you are to perform any cost benefit analysis, since this requires that you postulate 'before' and 'after' levels of risk and the corresponding cost. Expenditure typically secures an additional risk reduction factor (rather than an absolute amount of reduction) and the apparent benefit will therefore vary with the identified level of risk 'before'; there is more benefit from a factor ten reduction in the annual probability of a hazardous event from $10^{-2}$ to $10^{-3}$ than there is from $10^{-3}$ to $10^{-4}$.

It is all very well taking the anti-Panglossian[*] view of 'all for the worst in the worst of all possible worlds' but we need to be smarter than that when deciding how to allocate limited resources. If there are a number of provisions in the analysis that are known to be conservative by indefinite amounts, it may be appropriate to aggregate these and assign a factor to their combined influence in order to counter the distortion that would otherwise arise with compounding conservatism.

You may find that an iterative development of both your

---

[*] Pangloss is the character in Voltaire's *Candide* who continually asserts that 'all is for the best in the best of all possible worlds'.

understanding and the assessment is required until your increasingly educated expectations converge with the increasingly refined analysis. Far from being a 'massaging' of the numbers, this is the proper and responsible approach of the professional engineer. There is a world of difference between modifying a risk assessment because it is not judged sensible and because it does not fit the budget.

If it comes to a debate with the regulator about a risk assessment or safety case, remember that their inspectors will approach these questions from a distinct perspective. Understandably, this is unlikely to embrace your wider business concerns and resource constraints; it is for the duty holder to make a robust case that also recognises these issues. The regulator may well urge you to conservatism, which undoubtedly has its place in a prudent and responsible approach, but if you find this to be leading to unwarranted distortion, you should stand ready to defend your judgement.

# 3
# AND YET IT MOVES!

In 1633, Galileo was obliged to recant his view that the earth was not the fixed point at the centre of the universe, which was contrary to the teaching of the church. The Inquisition found him 'vehemently suspect of heresy' and he spent the rest of his life under house arrest. There is an apocryphal story that after recanting he muttered the phrase, 'Eppur si muove'/'And yet it moves'. It is unlikely that he actually said this, but we can be quite sure that he thought it.

At this remove it might, at first sight, appear as a quaint historical story, well removed from our modern 'sophisticated' culture. But there is nothing quaint about it. This was a highly significant episode in the history of science and it is representative of a commonplace dilemma for professional engineers in the 21$^{st}$ century: the engineering rationale points one way, the political pressure in another, which will win? Engineering or Politics? Be alert for this dilemma for it may come in many guises. Be prepared to jealously guard your engineering integrity! Be wary of siren calls to expediency.

Note that I do not insist on a refusal to compromise. There are many circumstances where compromises are an intelligent response to competing demands and expediency may be a valid consideration. It may be appropriate to compromise on the engineering execution in order to meet the wider project objectives. There is a legitimate debate to be had whenever the execution specifics conflict with the broader project objectives, for example, in terms of cost, deadlines and contractual obligations. Compromises on the engineering execution might well be appropriate, but it is a question of whether the proposed compromise is in pursuit of optimal project delivery or someone's personal agenda, whether declared or not.

Personal agendas might include a wish:

- To avoid responsibility
- To save face
- To secure a larger budget
- To acquire more influence
- To inflate importance

I do not say that all such objectives are necessarily suspect. It is when they conflict with good engineering that they become unworthy. Some engineers, as well as journalists, will not allow the facts to get in the way of a 'good story' (i.e. personal agenda). You might be urged to pursue one course over another, and that is fine as long as there is an honest debate about the merits of each. If you are ever invited to turn a blind eye or disregard engineering rationale however, then you must recognise this as a test of your professional integrity. Here, I may bend a line concerning honour from the film *Rob Roy* (1995) to my purpose: 'Professional integrity is what no man can give you and none can take away. Professional integrity is an engineer's gift to himself'. The corollary is that an engineer can only 'lose' his integrity if he chooses to give it away (or sell it).

Although typically there are levels of seniority and authority within engineering enterprises, most do not have a military style 'chain of command'. We are not required to execute orders unquestioningly; on the contrary, there is a professional obligation to challenge 'orders' if they do not make sense. (I intend no criticism of the military; their approach is necessary to their role.)

If an instruction does not make sense to you, there are only the following possibilities:

1. You have not understood
2. Your boss has not understood
3. Your boss has 'gone over to the dark side'

If the first, then you should seize the CPD (Continuing Professional

Development) opportunity. If the second, you should diplomatically highlight the CPD opportunity to your boss. If the third you should tread warily and look to defend your professional integrity. If the transgression is wilful and deliberate rather than inadvertent or misguided, then a formal complaint may be called for, but be sure of wrongful intent. Ultimately, if serious matters of safety or corruption are involved, the maintenance of your integrity requires that you blow the whistle long and hard.

More typically, you will meet issues of misalignment rather than outright corruption. Perhaps most insidious is the unthinking insistence on compliance with some tradition/standard/guidance where there is no consideration of the context, the underpinning engineering rationale, or the wider implications. Compliance is generally held to be a good thing, but if inappropriate, it will distort the engineering.

So what if you are being pressured to compromise the engineering for no worthy reason? It is all very well taking a stand but the personal consequences might be profound. Perhaps the best approach is to not vociferously protest (unless the transgression is blatant) but rather to summarise the considerations as a matter of record: without 'spin', without point scoring, without rancour - a simple summary statement of the engineering considerations with an acknowledgement of where judgement is required.

The point is that this will make apparent any misaligned behaviour; it will prevent distorted engineering being camouflaged with obfuscation. If others choose to misrepresent matters, that is their affair; if this leads to waste or inefficiency, well at least you did what you could to make the position clear. If you find yourself obliged to implement some distorted instruction or policy, then feel free to mutter, 'and yet it moves' and recognise that you take your place in a long history of denials in the face of evidence.

# 4
# PROFESSIONALISM AND THE ENGINEER

If you wish to claim the rights and privileges of a professional, you place yourself under certain obligations. *Noblesse oblige*. We might identify these 'rights' as a certain recognition and respect from others and a default presumption of your integrity; this is why registered engineers are recognised as persons of standing that may countersign passport applications, for example. Many of the obligations are codified in rules of conduct promulgated by professional engineering and other institutions. Typically, these institutions make it a condition of membership that you undertake to observe these rules. A key point however, is that your role as a professional is not confined to your professional role! The standards of behaviour apply to all business interactions with other individuals or corporate bodies. It is only in your intimate personal (non-business) relationships that the obligations might be relaxed.

The true professional (in the sense of belonging to a learned profession rather than that of being paid to do your job - an unfortunate confusion) cannot absolve himself from the associated obligations when he leaves the workplace for the day. You cannot sail under 'professionalism' as a flag of convenience that you haul down and replace with the Jolly Roger as and when the mood takes you. You cannot be partly professional any more than you can be slightly pregnant.

This burden might strike some as onerous, but it is the 'fee' for membership of the professionals' club. Fortunately, your membership is not immediately revoked if you should lapse in payment of your dues. (Which of us could claim never to misstep?) Any such lapses however, must be suitably infrequent, not systematic, without broader significance, with little profile and readily acknowledged. What is unacceptable is a wilful abandonment whenever the strictures become inconvenient. There are many aspects to professionalism, including:

- *Expertise/competence*: mastery of a specialised body of knowledge and a commitment to the maintenance of competence.

- *Responsibility/Accountability*: a willingness to accept responsibility for one's actions and to be held accountable. Although a layperson might claim, 'I was just doing what I was told', this excuse is denied the professional; the professional will acknowledge mistakes, not hide or deny them. He will not undertake work for which he is not competent.

- *Appearance/manner*: looking and acting the part - the maintenance of a professional demeanour. Speaking and acting in a measured way; considered and controlled. A becoming modesty; no aggressive self-promotion.

- *Courtesy*: due consideration and respect for others. This includes timely correspondence; 'every unanswered letter eventually answers itself'.

- *Self-regulation*: the avoidance of intemperate behaviour (think *Silent Witness* not *Eastenders*[*]).

- *Integrity*: honest dealing and honouring of commitments, duty before self-interest or inclination. This is at the core of professionalism. It is the foundation without which the other attributes become worthless. You cannot take it on and off like

---

[*] *Silent Witness* is a UK TV drama concerning forensic pathologists, *Eastenders* is a UK TV 'soap' drama.

a coat; you either have it or you don't.

The popular TV series *House* poses an interesting challenge to the notion of a 'professional'. Dr Gregory House is routinely rude to patients and colleagues; he dresses scruffily and breaks every administrative rule in the book. He does not suffer the mediocre (let alone fools), gladly. Although much of his behaviour is objectionable, the scriptwriters have contrived to make him an admirable figure because he is completely professional in this one respect: he has integrity. He is relentless in his pursuit of truth and logic, he refuses to compromise or fudge in that pursuit and, as a corollary, his patients benefit from his incisive diagnostic approach. It makes for compelling drama, but in the real world we must recognise that discourtesy would typically compromise our ability to effectively execute our professional role. To provoke this compromise would be unprofessional. Appearance *per se* is not intrinsic to professionalism but appropriate dress is a matter of courtesy; inappropriate dress implies a certain disrespect.

In a discussion concerning professional demeanour, the point was made to me that it would be unrealistic to expect professional software engineers to wear a suit and tie. That may be an unreasonable stereotypical judgement, but the key point is that dress should be 'appropriate'. The requirement is relative, not absolute. If jeans and a sweatshirt are normal in a given context then that is fine, but being perceived to 'cock a snook' would potentially introduce unnecessary impediments to effective interaction with others. It behoves us to adhere to perceived norms in dress for learned professionals if we wish to claim (and promote) the title, particularly when executing a professional role within sight of Joe Public.

Similar considerations apply to articulacy and literacy, but here the requirement is more absolute; an appropriate command of written and spoken language avoids ambiguity, misapprehension and potential confusion. Many aspects of professional engineering involve sophisticated provisions, nice distinctions and critical communications that can only be managed adequately with correct language. House may use slang, but he does so to add colour and spice to his speech. It is one tool in a wider box of articulacy (courtesy of his scriptwriters); he is not condemned to the use of slang by inarticulacy. (I have known the occasional swear word escape the lips of a professional engineer.)

Professionalism does not mean that you cannot break the administrative rules, but if you do, it should be for the greater good, not merely for some self-serving purpose. As Douglas Bader said, 'Rules are for the obedience of fools and the guidance of wise (professional) men'. However, we should of course be wary of straying into arrogance. I do not suggest House as a role model!

# 5
# BRING BACK THE SLIDE RULE?

The singular virtue of the slide rule was that it only allowed you to work to three significant figures and you had to work out where the decimal point went for yourself. The slide rule was an aid to calculation, but you certainly had to understand where you were going for it to be useful. There was also a certain tactile joy in the fluent manipulation of a slide rule. (Like a proper engineer!)

The crash in the slide rule market in the late seventies must be almost without parallel, the convenience and power of the pocket calculator was so completely overwhelming. Subsequently of course, the PC has swept all before it and there is previously unimaginable power within hand's reach on your desk or lap.

The undeniable power and convenience of the computer however, is not without its hazards, GIGO (Garbage In-Garbage Out) of course, but there are more subtle 'psychological' hazards. There is perhaps less tendency to question the result generated by a computer (a very necessary step in slide rule calculations!); it somehow appears authoritative. This appearance may be completely spurious though, remember the thing was programmed by a human being and the 'precision' of 10 decimal places is usually completely illusory. We are engaged in engineering, not physics; anything beyond the third significant figure is usually a joke.

At first thought, it is amazing that any human could beat a chess computer that is analysing gazillions of positions every second, but the computer still has to make an evaluation of the relative merits of each position examined. If the analysis does not reveal a forced checkmate or overwhelming gain of material, this evaluation must be based on 'positional judgement' that has been programmed by humans. The number of possible positions tends to explode astronomically with the number of future moves, so even this 'brute force' calculation has limited range. No Grandmaster can 'out calculate' a chess computer, but they may well display superior judgement of position that has developed intuitively and been honed by experience.

Generic assessment tools will typically be programmed with a degree of conservatism that may be completely inappropriate in particular circumstances - the program may well lack an 'is this sensible?' algorithm. (If I may be forgiven a moment's immodesty, a great deal of my value to my clients arises from running this same algorithm in my head.) Computers may be clever, but they often lack wisdom.

For many engineering purposes a spreadsheet is a perfectly suitable tool. When you are told you need something more sophisticated you should start to wonder. Refinement through sophisticated calculations might be proposed when the underlying foundations are necessarily simplistic and uncertain, in which case you have to consider whether the apparent refinement will offer real value; many a digital castle is built upon virtual sand. Of course, there are many circumstances where something other than a spreadsheet is appropriate; finite difference or dynamic response modelling calculations for instance, but many steady state calculations and database functions can be handled perfectly satisfactorily within a spreadsheet, which offers the advantage of a flexible and transparent implementation.

Computer tools may allow optimal design that trims the superfluous fat from your solutions, but might their use detract from the development of an intuition about what 'looks/feels right'? If used appropriately they may actually speed the development of such intuition by allowing the rapid assessment of alternatives by facilitating 'what if' games. If this philosophical point is not appreciated however, there is the danger that flawed designs might remain unchallenged by sound instincts based on experience.

This is where 'rules of thumb' come in 'handy' as a reality check on the computer result; if there is a significant discrepancy it may represent a useful refinement in the design, but you need to be sure you understand from what this arises.

> *...Let me give you one general piece of advice (...) Consider all structures, and all bodies, and all materials of foundations to be made of very elastic india-rubber, and proportion them so that they will stand up and keep their shape; you will by those means diminish greatly the required thickness: then add 50 per cent...* - Isambard Kingdom Brunel, 1854.

There is a lot to be said for having designers actually go into the field to commission their designs. Anyone with any experience will testify that what looks good 'on paper' doesn't always translate into corresponding operational experience. Even when the thing actually does what it was supposed to, there can be all sorts of headaches for the bloke that has to build it, or the operator that has to use it, or the guy that has to maintain it. Field commissioning will certainly hone your instincts for what works and what looks/feels right.

# 6
# GREY IS GOOD!

There is a lamentable reluctance on the part of some engineers to exercise their professional judgement, particularly in matters of health & safety. This often manifests itself in cries for 'black and white' standards. The problem here is that couching standards in completely unequivocal terms is likely to impose obligations that are inappropriate in many circumstances. If the stipulations were black and white, they would be 'rules' not 'guidance'. It is the role of a professional engineer, having acquired the appropriate competencies, to exercise professional judgement with due regard to pertinent guidance. We should rejoice in this. This is one mark of the professional - grey is good!

Attempts by some engineers to remove themselves from any question of liability can result in subtle but nonetheless significant distortions of business endeavours, with massive cost implications and even a net loss of safety. Professional codes of conduct do not say, 'Abdicate all responsibility where there may be a question of personal liability' or 'Distort engineering policy and practice as necessary to cover your behind'.

Good standards recognise the need to exercise judgement and in many areas will provide guidance rather than definitive stipulations. Some engineers will attempt complete compliance with a standard but compliance in every particular is often not a realistic ambition. You approach compliance asymptotically along a curve of diminishing return; you may approach closer and closer to full compliance but it requires ever increasing effort and investment. There is a point where the marginal increase in compliance does not warrant the additional effort, which may be more gainfully employed on other safety concerns. Professional judgement must be exercised to identify when this point has been reached.

Note that 'exercising judgement' does not mean the same as 'going out on a limb', which would imply accepting a significantly higher degree of risk; it will often simply come down to a matter of employing some common sense rather than 'going through the motions' of strict compliance even when there is little or no benefit (or even conceivably negative 'benefit'). Another misjudgement may be to attempt 'perfection' on a first pass; it may well be better to use a staged approach. If you first implement the more significant provisions, you may be better placed to make a more informed and effective approach for the aspects or areas that remain.

The wish of some engineers to avoid any accountability is recognised by some as a marketing opportunity. Occasionally you will find interested parties use 'scare tactics' to promote their product or service. It is all very well picking out individual clauses from the standards to 'demonstrate' a particular requirement, but without proper consideration of the context and the underpinning philosophy of the standard, it is easy to end up with a wrongheaded approach. If ever you are being told that you need to 'rip it all out and start again', whether metaphorically or literally, you need to take a step back and consider whether the 'logic' that is being deployed has carried you to a place that is no longer sensible. Do not fall into the trap of assuming a logical proposition is necessarily a sensible one, (which is itself an error in logic!). Careful consideration of the starting point and the explicit and implicit assumptions being made may reveal that a logical conclusion may be far from sensible.

I saw an example where a company had been hired to do a review of existing safety system provisions. A typical argument deployed in their report ran as follows (here distilled to its essence):

*A legacy system has a number of components. None of these has a performance level certified in accordance with a new (but not retrospective) standard. Therefore conservatively assume their performance level is $<X$. But the standard specifies performance level $>X$ for components being deployed in systems like the one being considered. Therefore replace all the components.*

Logical, but very far from sensible; the equipment had been procured and installed in accordance with good practice and there was no suggestion that it was not fit for purpose. An intelligent, responsible review, exercising appropriate judgement, would have identified this.

Unfortunately, some standards themselves suffer from this same condition: logical but not sensible. I can see how this comes about; you start with a simple premise and build logically, each step, each additional principle and consideration flowing inexorably, indisputably from there. The final structure is entirely self-consistent and rigorous but somewhere along the way it becomes so very much more than simply fit for purpose. You start with a brick and end up with a cathedral, when what you really needed was a granny annexe. Again, there is a need for standards committees to look back to the starting point and consider where logic has carried them and whether the position remains sensible in terms of the ultimate objective of the mission.

I do not suggest there is any conspiracy here, but there is a potential for experts to become seduced by their own expertise and to pursue excellence or refinement for its own sake or in competition with one another. There are also those that have a vested interest in elaboration, who look to become the new priesthood with exclusive rights to the performance of rituals. I am confident that it is possible to keep faith with many standards in a relatively straightforward manner, but keeping sight of the wood, when everywhere you look there are trees, can become a real challenge. Let me stress that it is not usually that the standards or guidance explicitly require over elaborate provisions but this is, I fear, often the inadvertent consequence.

The daunting prospect of some standards will understandably encourage many to seek support from consultants. The employment of consultants in this regard is entirely reasonable, but in attempting to button down every last aspect with complete rigour, in fulfilment of actual or perceived contractual obligations, there can be a tendency towards over-elaboration. I suggest that partnerships with consultants in this sort of development should focus on 'fitness-for-purpose' and 'prudence' rather than 'complete rigour'.

Another danger is that in order to try and place themselves beyond possible criticism, engineers will require all sorts of certified assurances in matters of supply/recruitment etc. The 'benefits' may be quite illusory and place unwarranted burdens and unduly restrictive constraints on business. Intelligent rather than slavish, unthinking compliance is needed. Regulators are faced with a complete spectrum of compliance within industry; from those that are well aware and strive to do the right thing, to those that are completely unaware and oblivious to the risks they run. Our predominant concern should be with the latter end of the spectrum, suitably weighted by potential consequence.

The important thing is to bring a considered, systematic and responsible approach to these matters; I cannot get too excited about whether every last 'i' is dotted and every 't' crossed (and I don't believe the regulators can either). Insistence on an entirely rigorous approach where not only the spirit of a standard is met but also every letter (including the 'i's and 't's) may well produce unwarranted distortion in deployment of resources, resulting in a net loss of safety. You need to consider where you will get the biggest safety 'bang'(!) for your buck. Consider also that there is often a trade-off between rigour and robustness; a more rigorous approach may look good on paper but may well prove more fragile in operation, with a tendency to suffer inadvertent corruption and with people prompted to make shortcuts (particularly when the collective management back is turned and operations continue at night or at weekends). This can in turn undermine the wider safety culture within an operation. As far as is practicable, a straightforward and relatively intuitive approach should be adopted.

# 7
# AN INSPECTOR CALLS...

Many find a visit from the regulatory authority inspector to be a daunting prospect, but let us try to establish some perspective here. Like any large organisation, the HSE (Health and Safety Executive) or the EA (Environment Agency) or their equivalents in jurisdictions beyond England & Wales have a spectrum of talent from the extremely able to the very much less so. Their inspectors are not demigods; they are idiosyncratic and fallible mortals. I have seen thoroughly professional inspectors, highly expert and experienced, who are prepared to take a pragmatic view of affairs. I have also seen inspectors who take an immediately hostile and pedantic stance, apparently with the idea of intimidating the inspected and thereby establishing ascendency over them. I have seen others who, uncertain of their ground, will bluster and make 'definitive' pronouncements that are in truth nothing of the sort. I have even seen one with a sense of humour (just joking!).

These organisations have their fair share of the egotistical, the insecure, the neurotic, the psychotic (just joking again!). None of this should come as a surprise; these foibles (and others) are to be found throughout our profession (and others). And, of course, a professional, competent, diligent, sane inspector can, like the rest of us, still get it wrong. The encounter need not (by default 'should not') be adversarial; the inspector can be your friend and can help inform and guide your activity. In this respect it may be useful to engage with him sooner rather than later. There should be a presumption of good faith from both parties and the default position should be one of mutual respect.

Naturally, you should extend every professional courtesy to the inspector, as indeed he should to you. This should be the case whenever professional parties meet. Whatever the niceties of conferred authority, if you wish to claim the title 'professional' you are bound by codes of professional conduct.

By virtue of his warrant, the inspector has powers that include the authority to:

- enter your premises at any reasonable time
- carry out investigations and examinations
- take measurements, photographs or samples
- require an area or machine to be left undisturbed
- seize, render harmless or destroy dangerous items
- require people to give accurate information or provide statements
- inspect and/or copy any relevant documents

These would typically only be invoked however, if there were an incident or want of cooperation. More usually, his attendance will be by appointment as part of a routine. Typically, if some shortcomings are identified, an inspector will take informal action with written or verbal advice. Only if there are more serious failures in provisions or timely action is he likely to formally issue an 'Improvement Notice' requiring you to take action in a specified time. If there is a more immediate and significant hazard, he may issue a formal 'Prohibition Notice' requiring you to cease a specified activity. The ultimate sanction is prosecution.

He cannot however, exercise these powers arbitrarily. If initial correspondence cannot resolve matters with your inspector and you believe there is an unacknowledged error or fault, you may complain to his manager. If formal notices have been served, you have the right of appeal. (An improvement notice is suspended pending the appeal, but a prohibition notice is not.)

Remember the inspector is a public servant, paid for by your taxes and the fees he levies. He has the 'luxury' of not having to manufacture anything; he does not have to juggle operational concerns that may conflict with safety issues. (Closing a plant will make it perfectly safe but there is a general recognition that there would typically be a net loss for society.)

This is of course as it should be. The inspector must be allowed to focus on safety. The implication is that you should not necessarily expect an immediate 'meeting of minds'. The inspector will have a distinct perspective on your affairs but you, not the inspector, are the expert on your process operation. If you believe the inspector's view to be inappropriate you should stand ready to challenge his position and deploy suitably rational and robust arguments in support of your own view. Be aware that a less able inspector may incline to simply reiterating the formal line rather than exercising professional judgement about what is practicable. Remember that it is very easy for the inspector to spend your money! (And he is unlikely to tell you that you are going over the top in your provisions.)

When I say 'challenge', I do not mean seek confrontation; rather it would typically be a question of offering alternative views or possible solutions. It should be a matter of informed debate between professionals, but if you believe unwarranted burdens are being dictated, you should be ready to protest. You have a right to expect a courteous and considered inspection and one would hope for good, even friendly relations, but remember that the inspector may need to maintain a certain professional distance. You may well need to investigate referenced standards and codes of practice yourself to identify just how pertinent their provisions are in the context of your operation. There may well be more than one way of meeting your obligations. I have known inspectors point to guidance that is not well suited to a duty holder's circumstances; the cited guidance or models might be useful in informing your deliberations, but should not necessarily be considered definitive and cannot necessarily be relied upon to provide the optimal approach.

It is not necessarily a question of having everything in a position to pass muster at first inspection, there may well be shortcomings in your provisions, but as long as the inspector can see that you are addressing these in a responsible and timely manner, he is likely to be satisfied and content to monitor your progress. Essentially the inspector will be looking for a considered, systematic, responsible approach informed by established good practice (including pertinent standards). If he can satisfy himself that you are on the right path, albeit not yet at your destination, he is likely to smile rather than frown on your endeavours.

Genuine misapprehensions are one thing, but attempts to deny shortcomings or obfuscate are quite another and run the risk of antagonising the inspector. Misapprehensions (from either party) should be readily acknowledged, not compounded by attempts at justification with a subsequent entrenching of positions. Attempts to deceive are, of course, unacceptable and contrary to professional codes of conduct.

I do not claim the considerations I have identified here as being solely due to observation of inspectors, they are in part compiled from the wider observation of professional engineers (or even the superset of human beings I have encountered) and much of what I have written will apply to any dialogue between professionals.

# 8
# CODES OF CONDUCT: STRAIT JACKETS OR SHIELDS?

Membership of many professional bodies is conditional upon acceptance of a professional code of conduct. As an example, a rule might require you to declare any conflict or potential conflict of interests. Now if you are found to be in violation of this rule, you have the unenviable choice of being perceived as a fool or a knave. The fool might declare, 'I did not know'; the knave would be seen as knowing and disreputable. What are the pros and cons here?

If you declare a conflict:
+ Compliant with the Code
+ Demonstrates integrity
- Risks loss of position/revenue?

If you don't declare:
- Violation of the code
- Risk of discovery with damage to reputation?
+ No immediate risk to position/revenue

So does a code of conduct represent a strait jacket or a shield? Certainly a code will act to constrain behaviour, but appropriate constraints are of course a requirement if one is to make any claims as a professional rather than some sort of rogue trader whose morals shift with opportunity. In terms of a shield, the code does afford a degree of protection for it can stiffen resolve, it can help resolve judgement calls and it can help by divorcing personality from a potentially unpopular stance; it is not 'you' that decides to disappoint rather you are obliged by the code of conduct imposed upon you. It may give you pause before, perhaps in all good faith, pursuing a seemingly expedient but professionally dubious course of action.

I am reminded of the following exchange in Robert Bolt's play, *A Man for All Seasons* concerning Sir Thomas More, Chancellor to King Henry VIII:

*Sir Thomas More has declined to employ the venal Richard Rich, who has just left the room.*

| | |
|---|---|
| **Wife:** | Arrest him! |
| **More:** | For what? |
| **Wife:** | He's dangerous! |
| **Roper:** | For all we know he's a spy! |
| **Daughter:** | Father, that man's bad! |
| **More:** | There's no law against that! |
| **Roper:** | There is God's law! |
| **More:** | Then let God arrest him! |
| **Wife:** | While you talk, he's gone! |
| **More:** | And go he should, if he were the Devil himself, until he broke the law! |
| **Roper:** | So, now you give the Devil the benefit of law! |
| **More:** | Yes! What would you do? Cut a great road through the law to get after the Devil? |
| **Roper:** | Yes, I'd cut down every law in England to do that! |
| **More:** | Oh? And when the last law was down, and the Devil turned 'round on you, where would you hide, Roper, the laws all being flat? This country is planted thick with laws, from coast to coast, Man's laws, not God's! And if you cut them down (and you're just the man to do it!), do you really think you could stand upright in the winds that would blow then? Yes, I'd give the Devil benefit of law, for my own safety's sake! |

Anytime you find yourself constrained by the code, this is likely to represent a positive opportunity to demonstrate your integrity by declaring the constraint to others. This can only enhance your reputation for professionalism.

There are occasions when commercial considerations might produce a professional and moral dilemma; for instance, when bidding for a contract you might be aware of some weaknesses in your company's offering, or of some likelihood of failing to meet a deadline. A full and frank declaration might damage your company's chances. Loss of the contract might mean redundancies (possibly including your own!) and real suffering for your colleagues, and yet non-disclosure might mean that the weaknesses later become apparent and contractual difficulties or ill-feelings may follow.

It is perhaps an uncomfortable line to walk but the line may in fact be quite broad; there is a big difference between wilfully misrepresenting a situation and casting things in a favourable light. It might also be that the weaknesses are reflected in your price or are compensated by other aspects of your offering. It must be acknowledged that some of these aspects are intrinsic to the nature of competitive tendering (*caveat emptor*). What you cannot do, and retain any claim of being a professional, is be dishonest. (I suppose that it is conceivable that a situation might arise where a lie is acceptable; provided it does not simply serve your interests but is for a greater good. This is seriously obscure territory, however.)

Perhaps the simplest and most robust test is that of propriety; given a proper understanding of the context, could your behaviour be publicly acknowledged without moral (as distinct from contractual) embarrassment, or damage to your reputation, or the wider reputation of the profession?

# 9
# TOUCHING WOOD: HOPE AND THE PROFESSIONAL ENGINEER

I am often dismayed by the sight of a professional engineer 'touching wood' after some declaration of a hoped-for outcome. Do they really believe that that is how the universe works? I trust not (or they have no right to the title and should have their epaulettes ripped off and their swords broken over a knee), but the prevalence of the gesture does demonstrate the insidious pervasiveness of such superstitions.

I believe we have a professional obligation to challenge such notions whenever we meet with them. (The continuing publication of horoscopes in mainstream newspapers gives the lie to the idea of an enlightened 21$^{st}$ century society.) I don't say we should pursue this line to the point of being objectionable but neither should we acquiesce without at least a good-humoured protest, *"All that is necessary for evil to triumph is that good men do nothing"* - Edmund Burke.

At first, it might be thought that hope has no place in professional engineering, but that is not correct; the important thing is that the professional engineer's hopes should not be blind. In troubleshooting, there may be a range of possible explanations for a problem or failure and we may legitimately hope for an explanation or root cause which has the least implications for rectification work and management. We might become aware that a component or system is degraded and approaching the end of its useful life, and we may legitimately hope for continuing operation until a convenient shutdown rather than suffer an interruption to production due to a breakdown.

Inevitably there will be a range of uncertainty associated with variables (or even 'constants'); we may legitimately hope for a value at a preferred point within this range, provided our hopes do not take us into unconscionably low confidence levels. What is not legitimate is to employ hope in lieu of evaluation, whether by analysis or the exercise of engineering judgement. It is not acceptable to make some design provision and simply hope it is good enough, or hope that some provision will remain viable without consideration of ageing/degradation mechanisms and diagnostic possibilities. Hope is not a legitimate basis of safety.

Engineering judgement is not simply the exercise of judgement in the field of engineering; it is judgement exercised on a rational basis with reference to identifiable engineering mechanisms and the employment of appropriate professional discipline. Note that I here deploy the term 'engineering mechanisms' in its widest sense of 'causal relationships' i.e. the relationships between causes and effects (not just mechanical assemblies with moving parts). If your judgements are not built on the foundations of credible mechanisms, they are not engineering judgements; they are guesses. I used the word 'identifiable' rather than 'identified' - the mechanisms may not always be explicitly identified, but they should always be identifiable. It may be that they are implicit in an intuitive evaluation or that they are understood as a 'given' by all concerned parties in a discussion.

The embedding of an understanding of engineering mechanisms is a key component in the education and development of a professional engineer. This allows us to deploy suitably honed intuition in our judgements. A judgement may be made intuitively but that is not to say that it should be made unconsciously. We should maintain an awareness of the relevant mechanisms underpinning our judgements even if we do not explicitly identify them.

These considerations underpin the intelligent use of standards and codes of practice in novel situations; it is only on the basis of an understanding of the relevant mechanisms that we can legitimately extrapolate/interpolate from established codes to new situations. I came across an example where an engineer had reverted in part to an earlier edition of a code on hazardous (explosive) area extent because the later edition was thought unduly burdensome in respect of provisions for pump seal failures. The argument deployed was as follows:

> *For a given pump, the new code stipulates radius R. The earlier code stipulated a hazard radius r. In terms of the <u>new</u> code, a radius r corresponds with a leak hole of diameter d. A hole of diameter d would produce more than the drips we <u>usually</u> see on the pump when a seal fails, therefore we can use radius r to define the hazardous area extent.*

What this did not recognise was that the provisions in the code were not based on 'usual' failures but on more significant but *rarer* failures (e.g. once in a hundred pump-years) with a correspondingly higher leakage rate and hazard zone radius. The new code provided for new radius R (>r) because research had identified that revised flammability limits should be used to determine the potential hazardous range for the fluids in question.

I don't say that the proposed provisions were dangerous (there is much uncertainty in determinations of extent and code provisions are typically conservative), but the point is that they were not traceable to any current code and nor were they substantiated by an appropriate mechanism to justify their adoption. The provisions were no longer coherent. (Continued use of the old code in all respects would at least have been coherent!)

We need also to consider the nicety of the judgement being made. The nicer the judgement, the greater the degree of rigour that should be employed in the identification of the relevant mechanisms and the greater the rigour with which we should seek independent validation. These degrees of rigour should of course also be weighted by the consequence of any error. Independent checks may seem superfluous where the influences of the mechanisms may be assessed in essentially black & white terms (i.e. not as nice distinctions), but the value of such checks may lie rather in making sure that all the relevant mechanisms have been recognised. (It may be that where nice judgements are to be made they tend to fall to senior engineers; there is perhaps a danger of professional arrogance here to be guarded against.)

By all means have hope, but don't use it as a design tool. Feel free to express your engineering hopes but please do not 'touch wood', especially in the sight of non-engineers; they might not realise you don't really mean it and you risk undermining the credibility of your profession. Trust me, I'm a professional engineer - the bridge won't fall down, touch wood!

# 10
# THE MEASURE OF AN ENGINEER?

There is a tediously recurring and futile lament in the UK about the lack of protection and status for the designation 'Engineer' and yet the title 'Chartered/Incorporated Engineer' is protected. It lies within our hands to endow these titles with the recognition we believe they deserve. How? Simply by insisting on their use in all appropriate circumstances and routinely deploying them in professional correspondence. I do not suggest we should routinely append a string of degrees and post-nominal affiliations to our signatures but a simple declaration of CEng/IEng as an indication of professional standing would not be out of place: Yours sincerely, F. Bloggs CEng. (It seems other professions are not so coy.) Nevertheless, let us consider just what the title does or should signify. This is perhaps of particular relevance given an increasing interest in questions of competence.

There are those who, somewhat naively, believe possession of a training certificate provides assurance of competence - but there is many a fool armed with a certificate. If the certificate relates to a specific craft skill and is suitably narrow in its application, then a certificate might offer suitable assurance. If the certificate relates to a truly professional field (in the sense of pertaining to a learned body of knowledge) however, then the possession of a certificate alone is unlikely to provide sufficient assurance. In the professional arena there is so much more to competence than the passing of an examination immediately following a training course.

## The Measure of an Engineer?

In seeking registration (as CEng/IEng), candidates must satisfy the criteria nominated in UK-SPEC (the UK Standard for Professional Engineering Competence) as promulgated by the Engineering Council UK, through a process of peer review. This includes a demonstration of UK&U (Underpinning Knowledge and Understanding). It is the establishment of this foundation that is critical. Specific knowledge is not critical to the professional competency of a registered engineer. Notional categories of competence such as trainee, supervised practitioner, practitioner, expert etc. do not so much constrain the scope of the work undertaken by the professional engineer, as indicate the lengths he should go to in researching and validating any eventual proposal he formulates before committing to a definite course of action. It is about knowing our prevailing limits, in particular when extrapolating from established ground.

We should turn to a professional engineer not necessarily expecting that he will provide an immediate answer (unless he happens to be an expert in that field) but rather that he will be able to identify and understand one. It is the ability to argue and validate from first principles and identify relevant mechanisms and rationales that is the hallmark of the true professional engineer, who is thereby equipped to exercise professional judgement and engage in self-directed CPD (Continuing Professional Development). Any registered engineer, by virtue of possessing UK&U, should be in a position to train himself through self-directed study of texts, attendance at conferences/seminars and consultation with peers etc. Given UK&U, we can anticipate that such CPD 'inputs' will correlate well with competency 'outputs'. It is the combination of UK&U and a commitment to CPD that constitutes professional competence.

Note that understanding is critical; knowledge alone is of limited use. It is understanding that allows us to extrapolate something new and useful. Feynman (our hero from Chapter 1) tells a story of how he teased his undergraduate classmates by telling them that the special thing about a French curve (a template for drawing smooth curves) was that it was made

> ...*so that at the lowest point on each curve, no matter how you turn it, the tangent is horizontal (...) They were all excited by this discovery - even though they had already gone through a certain amount of calculus and had already 'learned' that the derivative of the minimum of any curve is zero. They didn't*

*put two and two together. They didn't even know what they 'knew'. I don't know what's the matter with people: they don't learn by understanding; they learn by some other way - by rote, or something. Their knowledge is so fragile!*
[1]

Beyond UK&U, another area that might constrain professional competence is that of communication skills. The more esoteric the project, the more extensive in scope across a range of interested parties, the further from the well trodden, the greater the need for effective communication.

Some engineers ask, 'why should I bother with registration?' One of the key points is that it demonstrates that you have been evaluated by your peers and have been found to have acquired the appropriate competencies. If you are registered, in the absence of any contrary evidence, you may be presumed to be competent. Of course, it must be acknowledged that there are registered engineers whose professional competence may be in question and who are only fit to crank some well-established handle; the moment they go 'off-piste' they are likely to find themselves in difficulties. However, if the registration process is sufficiently rigorous, these instances should be quite exceptional, allowing our presumption of competence.

In assessing the professional competency of a job candidate, I would, in outline, propose the following questions:

- Is the candidate registered? (Not being registered does not imply incompetence, but no presumption of professional competence may be made.)

- If not registered, why not? (Particularly if the candidate is notionally eligible for registration.)

- If not registered, is the candidate a member of a professional body? (Membership may be separated from registration.) If not, why not? Is the candidate not sufficiently interested in his profession?

---

[1] Richard P. Feynman, *"Surely You're Joking, Mr. Feynman!": Adventures of a Curious Character* (London: Vintage, 1992).

- If not registered, is there explicit evidence of professional competence?

- If registered, is there evidence contrary to the presumption of professional competence?

- Is there evidence of self-directed CPD? A list of courses that the candidate has been 'sent on' would not constitute evidence.

- Does the candidate have the appropriate communication skills?

I would not put on the list, 'Does he hold a related training certificate?' nor 'Does he have relevant experience?' These questions do not address professional competence but rather 'role-specific competence'. The engineer who is both professionally competent and experienced will likely identify appropriate solutions more quickly than his professionally competent but inexperienced colleague. If we simply see competence as a question of what training certificates the candidate can wave, we will have very seriously missed the point. I would not see a past mistake as contrary evidence, unless the mistake was systemic. Indeed you might argue that it would enhance competence if the associated CPD opportunity has been embraced.

In the revision to the functional safety standard BS EN 61508 (now $2^{nd}$ Ed.), the requirement for competency has been changed from being 'informative' to 'normative' meaning that it is now mandatory if compliance to the standard is to be claimed. Much has been made of this and many are pursuing certification as 'Functional Safety Practitioners', but possession of such a certificate does not demonstrate competence, it demonstrates training. (I have seen certified practitioners that I would not pay with buttons.) Consider certified training as one possibly useful element of CPD but not necessarily to be preferred over self-directed study and relevant experience. Certified training, in the absence of professional competence, is virtually worthless. If combined with professional competence, it offers the beginnings of role-specific competence but it is of limited value without experience.

Regardless of whether designated 'informative' or 'normative' within a standard, the requirement for competence has always been there. If a professional engineer chooses to act in a responsible capacity without appropriate competence, he opens himself to a charge of negligence; this has always been the case. Some engineers have a tendency to 'abdicate responsibility' in an attempt to avoid any question of liability. This manoeuvre may help reduce the personal risk of a charge of negligence from a regulator, but it should invite a similar charge from the employer - dereliction of duty, want of professionalism, fraudulent posturing as a professional engineer and negligence with company resources. In short: wilful incompetence.

The measure of a professional engineer? Not whether he knows the answers, but whether he knows and understands the questions - and is prepared to ask them.

# 11
# SAFETY MANAGEMENT: BEWARE THE REALITY GAP

The question is: how best to address the increasing burden of safety regulations? One approach is to throw money at the problem until it (apparently) goes away. This is likely to involve extensive use of consultants and contractors. The 'apparently' proviso is prompted by the possibility that the external resources used may not be entirely competent, or may, even in all good faith, introduce measures that are inconsistent with the client's circumstances. There is a danger here of 'management by abdication', which is in effect no management at all. It should be recognised that consultants do not have the same motivation to keep things as simple as possible. There is also the risk that your own people may 'over egg the pudding' in order to further their own standing within the company.

Many an empire or reputation has been built on somewhat spurious grounds. It should also be recognised that unquestioning, wholesale implementation of standards may inadvertently increase risk in some respects. The introduction of impractical measures may weaken the wider credibility of management systems and lead to poor compliance, or divert resources from areas where there may be more pressing needs that might give a better return in terms of safety. If a 'reality gap' is perceived between what we say and what we do, or what we do and what we need, the whole safety culture within an organisation is undermined. Safety measures must be understood to be commensurate with the risks if credibility is to be maintained. Better not to issue a directive if you are not minded to ensure it is implemented. Many are tempted to issue a procedure or instruction or put up a sign, without any real commitment to effective implementation, presumably in the belief that this will provide a defence against any charge of negligence. That it will, but not one that is likely to stand if accepted custom and practice allows routine violation.

I distinguish here between regulations that must be complied with if an operation is to remain on the right side of the law and which the regulatory authorities have a duty to uphold and standards where compliance is not mandatory but is said to represent good practice. The regulatory authorities would do themselves a disservice if they failed to recognise the realities of individual circumstances and simply insisted on unthinking compliance. The law of diminishing returns applies and, besides the possibility of tying up resources that could be better directed, there is the possibility that strict compliance may produce a less robust solution that is more susceptible to breakdown.

There is usually a trade-off between rigour and robustness in administrative practice. The optimal solution that produces the best overall safety performance may not be the same as one that is nominally entirely compliant. That is not, of course, to say that inconvenient or difficult provisions should be disregarded; rather it is provisions of marginal benefit that require unwarranted resources that should be challenged. The test is whether it is foreseeable that any deviation could significantly contribute to the likelihood or severity of an incident. The need is for a suitably pragmatic approach. This may well mean that implementation is not completely compliant in every particular with regard to a given standard or set of guidelines. The standards are normally generic in nature, in order that they may have the widest applicability; some interpretation and intelligent discrimination is often called for.

Most modern regulatory developments are based on a risk assessment of potential hazards and it is appropriate to extend this approach to implementation of the standards themselves. The standards are usually written in 'legalese' to avoid any ambiguity and prevent the irresponsible from wilful misinterpretation that complies with the letter but not the spirit of the content. An unfortunate side effect is that it discourages flexible but responsible interpretation by those anxious to do the decent thing. It is the user who must come to a considered view of what is appropriate. If any external party is consulted, it is almost inevitable that they will simply reiterate the formal line. A good example of where unwarranted rigour may actually compromise safety is in the management of change. An entirely rigorous approach may look good on paper but if it is overly cumbersome, compliance, particularly out of hours e.g. nights and weekends, may be poor. Systems that are perceived to be unduly burdensome actually promote shortcuts and may well result in a net loss of safety.

There is a danger that safety management may preach such systems from an 'ivory tower', thinking that they have thereby demonstrably discharged their responsibilities, but a failure to address any reality gap may mean safety is actually compromised by the 'improvement'. It is important that safety management be effectively integrated into the overall business management to the extent that it shares ownership of the problem as well as the solution. Equally, production, maintenance and design management must recognise their responsibility for closing the reality gap from their end. The challenge for those in engineering management is to identify a considered, systematic approach effectively tailored for the specific circumstances. This takes real skill, insight and diligence; it is also what makes the job interesting. Simply quoting the rulebook is an abdication of professional responsibility.

# 12
# ENGINEERING, FAST AND SLOW

In his bestselling book, *Thinking, Fast and Slow*[1], Daniel Kahneman explains some of the flawed mechanisms that can influence our thinking. He points to two distinct systems of thinking: 'System 1' (intuitive, automatic) and 'System 2' (deliberate, controlled and rational). These are the 'Fast' and 'Slow' of his title.

He discusses the nature of expert intuition and reports that the factors that point to a reliable intuitive judgement from an expert are:

a) 'an environment that is sufficiently regular to be predictable' and
b) 'an opportunity to learn these regularities through prolonged practice'.

He makes the key point that the confidence of the person proclaiming a judgement is a very unreliable guide to validity and should be disregarded. Consider rather, whether the field they are pronouncing upon is regular and whether they have the necessary experience.

Certainly, we may class engineering as a 'regular environment' but there is still some difficulty in assessing whether someone has the necessary experience. Note also that experience is a necessary but not sufficient condition for competence. There are those that cannot

---

[1] Daniel Kahneman, *Thinking, Fast and Slow* (London: Penguin Books, 2012).

effectively generalise from specific experience (a key attribute of intelligence) and individuals will differ in the amount of experience they need to reach a given level of competence. 'Sufficient experience' will also be a function of the context and the complexity/subtlety of the question being addressed. Experience (in the unqualified sense of time served) may even be a bad thing if it is not related to good practice. As long as it is understood for what it is however, a bad experience can be good experience!

So how are we to assess whether someone, ourselves included, has sufficient experience to offer sound judgements? It is largely a matter of judgement! We appear to be faced with a recursive approach. Not that we need despair however, recursive approaches may converge on useful solutions. The variables of the experience 'function' are extent and relevance; the parameters (characterising the individual) are UK&U (Underpinning Knowledge and Understanding) and we can validate our assessment of competence by continual cross-checking of judgements with our peer group and its publications.

Without UK&U, exposures to all the 'relevant extent' imaginable will still not constitute sufficient experience. It is understandable that demonstration of UK&U is a key requirement for those seeking registration as professional engineers. It is this attribute that provides the understanding of the causal mechanisms that regulate the engineering environment.

When training engineers, I am sometimes dismayed to encounter a willingness to merely crank the handle on a sausage machine without any attempt to understand the mechanism/model (often embedded in software) that produces the results/sausages. We cannot allow basic mathematical techniques and concepts to be regarded as abstruse (and quickly forgotten) hurdles, perversely placed in the student's path to qualification as a professional engineer!

A good engineer will always assess the 'first derivative' of his judgement; in other words the sensitivity of his judgement to assumptions and case specific estimates:

$$\delta judgement \approx \frac{\partial judgement}{\partial data}.\delta data + \frac{\partial judgement}{\partial model}.\delta model$$

Proper judgement will always depend upon two categories of information: the available data and a model of the situation. Very often the model will not be an explicitly mathematical one, it might very well be 'fuzzy' and intuitive, but any engineering judgement must be based on a relationship between pertinent variables and parameters.

I encountered an example recently where it was proposed that a safety case should be assessed on the basis (in the event of storage tank failure) of 70% of the contents overtopping the associated bund of 110% capacity. I was surprised by this 70% figure, which I could not square with my intuition. It turns out that the figure arose from a report[2] of physical scale modelling of near instantaneous disappearance of a tank (representing catastrophic tank failure), leaving a column of water free to collapse under gravity in a near frictionless environment (acrylic sheeting). The consequent tsunami carried the majority of the contents over the bund. The report identified up to 70% would overtop the bund, the estimate of the amount depending on the ratio of fill height to bund height and the tank proportions. The near instantaneous tank disappearance and near frictionless environment can be seen to constitute a bounding case - as claimed in the study.

I do not question the findings, the study methods or the rigour with which the work was done or reported, but what was not addressed was the 'first derivative' of the findings - no doubt beyond the study brief. The target time for lifting the model tank wall clear of half the column height was less than 0.1 seconds. The question arises: what would the results be if the target time was 0.2 seconds or 0.3 seconds? Even with catastrophic failure, I can imagine the tank offering some resistance to outflow. How representative is the 0.1 second target? How sensitive to this timing are the results? Possibly the results would be substantially the same if the time was doubled or trebled, but how are we to judge? A bounding case may well have been faithfully established but if the boundary is very far removed from the reality, it may represent an unduly onerous condition for risk assessment purposes.

A separate scale model study[3] conducted three years earlier reported

---

[2] William Atherton, *"An Experimental Investigation of Bund Wall Overtopping and Dyanamic Pressures on the Bund Wall following Catastrophic Failure of a Storage Vessel" Research Report 333*, prepared by Liverpool John Moores University for the Health and Safety Executive (Surrey: HSE, 2005).

[3] P. S. Cronin, J. A. Evans, *"A Series of Experiments to Study the Spreading of Liquid Pools with Different Bund Arrangements" Contract Research Report 405/2002*, prepared by Advantica Technologies Ltd. for the Health and

overtopping fractions of approximately 1 to 10% (depending on bund geometry, bund wall slope and tank fill height) in response to a full contents discharge from a tank base in 30 seconds. I do not claim expertise in these matters; my intuition may well be wrong, but it is the starting point for some critical questioning of the models being used. Without some indication of the sensitivity of the model to the assumptions made, my doubts will remain. Even if the model is inherently sound, the question arises about how well it reflects the circumstances; is catastrophic failure of the type being modelled a credible scenario for the tank in question? Modern tanks are typically built to codes that reduce the risk of catastrophic failure due to internal explosions or brittle fracture. Older tanks may be susceptible to these mechanisms and, for all I know, may be effectively modelled by virtually instantaneous tank disappearance.

Most engineers recognise that 'Garbage In' means 'Garbage Out' (GIGO), but they are perhaps not always as alert to the possibility of 'Quality In, Garbage Out' (QuIGO?) arising when a flawed or inappropriate model is used. These considerations point to one of the key requirements placed on professional engineers: know your limits. One of the fundamental provisions in the Codes of Conduct that professional engineering institutions require their members to abide by is that of not undertaking work for which you are not competent, but how are we to know our limits? There is no difficulty when the issue is far removed from our normal field or range of experience; but when it is a marginal case, it is far from straightforward.

This is where CPD (Continuing Professional Development) is important; it is not just about acquiring competence, it is also about not losing it and avoiding obsolescence. It might be better to call it MPC (Maintaining Professional Competence). It is through continual validation, by cross-checking with our peers and their publications, that we maintain an understanding of the range of our own competence. It is often assumed to be a matter of attending a course/conference here or there (and certainly these can be useful components of MPC) but more fundamentally, it is about maintaining an interest and being open to information and debate in your chosen field. It is in facilitating this that the professional engineering institutions find a *raison d'être*. Recognise also that just because you are

---

Safety Executive (Surrey: HSE, 2002).

competent in the engineering context, does not necessarily make you competent to make judgements with respect to associated commercial or ethical questions concerning development or deployment that may arise in a wider context. You will likely be in a position to offer a considered opinion, but that should not be confused with competent judgement.

We should not make the mistake of thinking that CPD is only for the young aspiring engineer; there is a real threat of creeping obsolescence in our knowledge (hence my shift in emphasis to MPC). If this is coupled with an arrogance that might come with seniority or maturity, there is a danger of invoking authority and overlooking or suppressing contrary views. It is one thing to invoke your authority to suppress spurious debate and 'noise', quite another to squash probing discussion. If ever you catch yourself playing the equivalent of the 'Do you know who I am?' card you should realise that you have crossed a line somewhere. Recognise also that the suppression may be completely inadvertent, for example, an entirely justified air of confidence in a senior figure may discourage qualifying views or questions. A conscious effort may be needed to mentor contributions from junior colleagues, which may be found to add unanticipated value - and junior engineers should resolve to ask the questions regardless.

It takes a degree of courage to ask what you fear may be a stupid question but 'stupid' questions often turn out to be not so stupid after all; even if they are not of direct concern, they may well prompt consideration of things that might otherwise have been overlooked. Senior engineers who have migrated into management positions may be particularly susceptible to these effects and should stand ready to acknowledge that those they manage may well be more competent in some of the functions being managed.

I was struck by the wisdom of the following extract from the introduction to the book, *Judge Sewall's Apology*[2] (Sewall was one of the judges in the Salem witch trials of 1692, who subsequently made a public apology for his role in the affair):

> ...*For all of us, it's difficult to say sorry. An apology means repudiating an aspect of our past selves; in a way it's like a little suicide. And those in public*

---

[2] Richard Francis, *Judge Sewall's Apology and the Forming of a Conscience* (London: HarperCollins, 2005).

*life find it almost impossible ever to admit they have made mistakes and errors of judgement. They fear doing so will suggest weakness and unreliability, a poor capacity for decision-making, ultimately a fatal crack in the façade of leadership....But there is another way of looking at it. Apology can be a creative act. It can liberate both an individual and his or her society. Apology frees you from the past and gives access to the future. It allows a person to evolve in response to the shifting demands of life. It requires an ability to inspect and evaluate your actions as if from the outside, and demands a high degree of courage and honesty. Apology can be heroic...*

There are clear parallels with the need for the professional engineer to acknowledge errors and misjudgements and, if circumstances should reveal a weakness, to be willing to embrace the MPC opportunity.

# 13
# RISK ASSESSMENT AND FERMI CALCULATIONS

We are routinely enjoined to manage risk to be ALARP (As Low As Reasonably Practicable) and given the legal requirement in this respect (in the UK), there is an understandable interest in identifying just where ALARP lies in any given context. There is the potential here to start all sorts of spurious hares running and we need to be careful not to tie ourselves up in unwarranted knots.

Perhaps a good (?) example is the Energy Institute's, *"Guidelines for Managing Inspection of Ex Electrical Equipment Ignition Risk in Support of IEC 60079-17"* published in 2008. The document is an impressive piece of work that appears consistent and logical; the determination of a sample inspection methodology on a rigorous statistical basis is undoubtedly clever but given the nature of the inspection challenge in practice, I find the proposed approach to be fantastically over-elaborate and completely at odds with the broad uncertainties in this area. These guidelines purport to offer a sampling methodology that is ALARP and yet they explicitly acknowledge that, '...the sampling methodology does not account for the consequences of a fire/explosion, nor does it take any credit for any mitigation measures'.

Now call me peculiar, but any health and safety risk assessment that does not examine consequence strikes me as being of questionable value. In promoting what I believe to be over-elaborate protocols for the management of inspections, these guidelines may do industry a serious disservice, particularly if they are perceived to constitute any sort of benchmark of good practice.

Let us perform a quick Fermi calculation here, named after the physicist Enrico Fermi (who apparently often performed such calculations with surprising accuracy), which is also sometimes referred to as a 'back-of-the envelope' calculation.

The UK's HSE (Health & Safety Executive) identifies an average annual fatal accident rate (for the five years to 2008/9) in the UK manufacturing and utility supply and extraction industries of approximately 5 per 100,000 workers. Let us assume, with massive conservatism, that these fatalities are all due to unrevealed degradation of 'Ex' (Explosion Protected) equipment items operating in hazardous areas (more typically they are due to falls from height, vehicle movements and accidents involving machinery). Remember also that hazardous area provisions are not intended as a defence against catastrophic plant failure and very large releases. 5 fatalities per 100,000 workers per year is a 5E-5 fatal risk per year for each worker. Now the cost of inspection will of course vary with the size and nature of any given installation, but we may normalise our analysis as follows.

The individual worker is not exposed to all 'Ex' release hazards simultaneously; he will typically be within range of a limited number of release sources for a portion of his working day. Let us assume however, that he is continuously exposed to an explosion hazard as he moves about his site. If there is a 10m radius zone from any given release source, then all 'Ex' equipment items within that radius represent a potential source of ignition if they have suffered degradation that has not been addressed by inspection.

Let us assume that 50 equipment items within the 10m radius constitute the hazard to be addressed by inspection. (It does not matter whether it is a small site with just 1 worker continuously exposed to the risk associated with the 50 items, or whether it is a large site with 5,000 items with 100 workers, each spending 1% of his time exposed to each of 100 individual groups of 50 items.) The cost of inspection, we may conservatively estimate as 10 items per hour at £30/h, giving £150 per group. If this is every three years, we have an annual inspection cost of £50 per worker against the aggregate hazard to which the worker is exposed. So we spend £50 per worker per year and tolerate an individual worker risk of 5E-5 fatalities per year.

We may postulate that the inspection cost will be proportional to inspection frequency and that risk of fatality will vary inversely with inspection frequency. So if we double inspection frequency we would double the cost and halve the risk. The incremental cost/benefit of doubling inspection frequency would be an additional £50 per worker per year with an additional 2.5E-5 lives saved per worker per year or £2M per additional statistical life per year.

Given the massive conservatism in the analysis, this demonstrates that existing 'Ex' inspection practice within these industries is already ALARP. A comparable analysis could be performed in matters other than inspection of 'Ex' equipment. What this essentially demonstrates is that the process industries have been well served by established good practices and that there is no pressing need to attempt more rigorous determination of the basis of inspection. Only exceptionally would I expect to undertake an explicit incremental cost/benefit calculation, typically in respect of a specific high hazard scenario. The higher the risk, the greater the requirement for rigour in the demonstration of ALARP and the more confident we need to be that uncertainties in the demonstration will not result in operation in the intolerable region.

Whatever the outcome of any CBA (Cost Benefit Analysis), there is an overriding requirement to adopt prevailing good practice. Note that I say 'good' practice not 'best'. There may always be someone who does it better, but that does not mean that this represents a model practice to be adopted; 'best' is a superlative that implies 'exceptional' and would therefore only be warranted in an exceptional context. Standards and codes of practice will likely help identify good practice but they often present this from the ideal perspective of starting with a clean slate. The real world user may have to find a pragmatic compromise. Standards may include refinements that may be of marginal value in a given context. It is a poor engineer (or inspector) who simply reiterates the formal line without exercising professional judgement about the significance of any deviation.

An intelligent review will recognise that there are provisions that, although 'nice to have', are not critical to responsible risk management. Typically, accidents arise from a failure to observe the fundamentals of good practice rather than any failure to dot 'i's and cross 't's. Standards such as those associated with functional safety (IEC 61508/61511) are so comprehensive in scope and so detailed in provisions that absolute compliance is hardly to be believed or required.

We might deploy a similar argument to the Fermi calculation above in respect of functional safety and thereby demonstrate that (in aggregate at least) existing functional safety provisions are already ALARP. There is no surprise here; the standards were not introduced due to a wave of fatalities from functional safety failings, although you could be forgiven for thinking that on seeing the extent of these standards. The core of these standards is perfectly sensible but attempts to implement this on an entirely rigorous basis, without recourse to the exercise of professional judgement, will likely lead to unwarranted difficulty and undue elaboration.

The safety practitioner community has something of a self-serving tendency to pursue increasing elaboration in the approaches it advocates, regardless of whether it is really warranted. As elaboration is held to represent a notional improvement in safety, it often goes unchallenged, but any proposed elaboration or refinement of approach should itself be subject to cost/benefit analysis. A quick Fermi calculation may demonstrate that the 'improvement' is not justified. Safety practitioner pronouncements tend to focus on the ideals rather than on more practicable approaches that, although less than entirely rigorous, may still effectively address the key concerns. Safety is often played as a trump card, defeating all other cards or considerations on the table regardless of their value, but this is wrong; the safety argument should win the trick on merit, not because of some misplaced notion of political correctness.

# 14
# STAYING ALIGNED

In Chapter 10 I considered the potential for a disconnection to arise between requirements and implementation. This potential exists in all disciplines of course, not just that of safety management. Let us consider some of the characters that we may meet who will inadvertently or otherwise lead to the development of a reality gap because their efforts are poorly aligned with the business objectives:

- *The Expert:* will seek to exercise his expertise at every opportunity, regardless of the business requirements and will seize every opportunity to enhance or promote his expertise.

- *The Enthusiast* (closely related to the Expert): will pursue his 'hobby' at every opportunity and will promote wider involvement with his enthusiasm as he seeks to evangelise. He differs from The Expert in that he does not jealously guard his knowledge.

- *Fashion Victim:* is seduced by the latest fad without proper consideration of its true relevance to the business endeavour.

- *Empire Builder:* his efforts are primarily directed to building his power base and influence. The business is simply a convenient vehicle.

- *Cover-My-Ass Brigade:* these guys avoid making judgement calls. They will always adopt the most conservative option, regardless of profit implications and will call in a consultant at the drop of a hat, (which is not necessarily a bad thing providing the consultant is efficient in execution of the assignment and delivers on a brief that is well-aligned with the business objectives. If he borrows your watch to tell you the time, you will only have generated apparent profit).

- *The Fear Monger:* will (spuriously) prophesise disaster unless their recommendations are adopted.

- *Headless Chicken:* is locked in panic mode and will promote an over-the-top, knee-jerk response. (Not that they will describe it that way of course.)

- *The Ostrich:* will pretend the problem does not exist in the hope that it will go away. It can happen, but I cannot recommend it as a business management approach.

- *The Knight* (in shining armour): will pursue what he perceives to be a noble cause regardless of the business context. He is a useful guy if you happen to have a damsel in distress, but if your business is relatively distressed damsel free, he is likely to trip up your business with his misdirected lance.

- *The Deckchair Attendant:* is penny wise and pound foolish. He will distort the allocation of resources and energy to salvage relatively trivial sums without considering the wider impact on the business. (This is the guy that would redeploy a look-out to rearrange the deck chairs on the Titanic.)

- *The Extra-Terrestrial:* is from a different planet to the rest of us and has lost the plot completely. He sometimes has harmless entertainment value, but if in a position of authority, he is likely to start all sorts of spurious hares running, which take valuable resources to trap.

- *The Social Worker:* is so concerned that we should all stay friends and be fulfilled that the business can go hang in the meantime. (Sometimes confrontation and anger are forces for good, particularly if you happen across blatant, wilful distortion of your business.)

- *The Mystic:* has supreme faith that all will, in some mystical way, come good in the end because this is the deserved, right and proper outcome. I distinguish here between the blind faith of the mystic and faith expressed as a professional judgement based on a plausible causal mechanism.

All of these guys (or gals) are particularly dangerous when they are armed with a budget. Nothing exhaustive or rigorous about any of this of course, and the disturbing truth is that we are all capable of exhibiting many of these behaviours to a greater or lesser degree at times. If we think of 'profit' here, not as a financial measure, but as the wider benefits accruing to an enterprise from the efforts expended, we can conceive of a vector quantity for resource allocation or expended effort and consider its alignment with the true business objectives. It is only the (Cos$\Phi$ x Expended Effort) component that generates real profit.

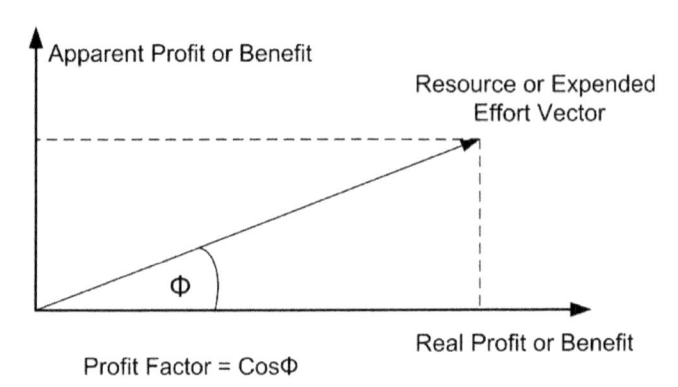

Apparent Profit or Benefit

Resource or Expended Effort Vector

$\Phi$

Real Profit or Benefit

Profit Factor = Cos$\Phi$

Any effort that is not properly aligned with the business objectives will generate 'apparent' benefit or profit as distinct from 'real' benefit. The classic manifestation of this is the 'busy fool' who works flat out to no good business purpose. (Electrical engineers will appreciate the parallels with apparent power associated with reactive loads.)

The fundamental thing is to acknowledge that a significant proportion of the effort expended in a business may be improperly aligned with the business objectives. I do not claim that this is necessarily deliberate. Much of it is inadvertent; people will typically strive to do the right thing, but unperceived and subtle distortions may well arise because of misunderstandings or a failure to properly appreciate the true business context of their efforts. Some distortions will be deliberate (but not necessarily self-serving) as people pursue their personal agendas because they believe the management-nominated direction is flawed in some way. Of course some misdirection will be both deliberate and self-serving as people pursue personal ambitions.

Perhaps the happiest circumstance is when a direction is both self-serving and fully aligned with the true business objectives. In engineering there is often a pursuit of excellence for its own sake, but 'best engineering practice' is not necessarily 'best business fit'. We need to consider the practice in terms of the particular business context we are engaged upon. Don't take your business hat off when you put your engineering hat on. The need is for a rational, prudent approach, which tempers the pursuit of excellence with a pragmatic recognition of the operational context and the business objectives.

We often hear success stories relating to implementation of different gurus' approaches to revitalising business. Of the returns typically identified, how much is due to the fact of having spent the time and energy in investigation and how much is due to the specific methodology employed? I suggest that pretty much any sensible, systematic approach would deliver the majority of the returns if applied with appropriate time and energy. What is the role of the professional engineer if not to bring a sensible systematic approach to business improvement and management? For the most part of course, it is not the intrinsic virtues of the approach that yields the results; it is the motivation, focus and enthusiasm engendered in the 'converts'. Hey, if it works don't knock it, but be clear about what you are doing and why.

So how do you avoid misalignment? I'm afraid there is no equivalent of the electrical power factor correction capacitor: you certainly don't want people pulling in different but balanced directions so that the net resultant resource vector is properly aligned. The requirement is for continual vigilance, the maintenance of a healthy scepticism and recognition of the problem and a willingness to tackle it. Stay alert. Develop your instinct for misalignment. Routinely take a step back from the fray and ask yourself, 'What are we trying to achieve here?' It is a simple but nonetheless profound question. Once you start to critically review this, you may be surprised at how often people fail to ask this basic question and then set about a problem that is more interesting, more tractable, more comfortable, but imperfectly aligned with your true business aims.

# 15
# JANE AUSTEN: ENGINEER?

I am pleased to declare that I am an admirer of Jane Austen. (In a moment of flippancy I did think to say 'confess' but that would carry a connotation of shamefulness which is entirely absent.) There are those that would dismiss her literary canon as 'chick lit' but we need not concern ourselves with these barbarians, who are unlikely to make it past the title of my essay.

The genius of Jane Austen lies in 'the most thorough knowledge of human nature, the happiest delineation of its varieties, the liveliest effusion of wit and humour (…) conveyed to the world in the best chosen language', as one of her characters says of novels as an art form. The particular pleasure to be derived from her work lies chiefly in the intelligence and wit with which she draws the nicest of distinctions; only if it were merely a matter of regency elegance and a study of manners might it deserve the appellation 'chick lit'. Let us consider however, whether Jane Austen had the necessary attributes to be an engineer - assuredly a piece of whimsy, but I enjoy the sport and might in the process establish some insights into the nature of these attributes.

Clearly, Jane possessed a remarkably fine intelligence. I conjecture that such intelligence would readily have absorbed the necessary learning in matters of physical laws and properties. I suppose there is the possibility that she was dyscalculic, which would perhaps have disbarred her from the heights of our profession, but I see no evidence for this or any other reason to doubt she could have acquired the required UK&U (Underpinning Knowledge and Understanding).

I trust the discerning reader will not need much by way of *persuasion* that Jane would not exhibit *pride and prejudice, nor th' anger* associated with an intemperate nature but would remain objective and clear-sighted. Engineering is often considered a *man's field*, largely I suspect because it is seen as requiring *sense* rather than *sensibility*. Certainly, as professionals we should seek to *emma*-ulate (ouch!) models of rationality and propriety but it would be wrong to suggest that there is no place for sensibility.

Between the plain wrong and unworkable, and the indisputably correct and effective, there is a world of possibilities where the identification of the appropriate solution requires not only sense (to establish viability) but also sensibility to aesthetics and to context, circumstance and culture. Jane's novels demonstrate both sense and sensibility in abundance. Her particular talent was an ability to identify subtle yet profound distinctions and exercise proper judgement; not unthinkingly on the basis of a code (of etiquette) but from an understanding of first principles (of morality and humanity). I am confident she could have extended this approach to matters of engineering judgement.

There are those that see professional engineering as nothing more than the competent cranking of a handle on an appropriate mechanism (whether physical or procedural) and look to execute their allotted roles in this manner. Clearly, there is a need for competent handle cranking, in engineering as with any profession, but such a pedestrian view will not give rise to originality or intelligent and discerning execution. It is, of course, possible to competently crank a handle to no good purpose or to one that is poorly aligned with the proper objective. Jane's novels are not formulaic; certainly, her books inhabit relatively narrow territory, but they demonstrate a remarkable range of tone (the sombreness of *Mansfield Park* is in notable contrast to the gaiety of *Emma*). Jane recognises the boundaries of her competence however, and does not stray.

Jane certainly demonstrated a mastery of language - something that some engineers affect to disdain, perhaps because they confuse poetry (as a celebrated but sometimes oblique form of language) and precision. If we claim to value precision in our profession though, we must logically extend this to our personal communications.

As a profession, I believe we like to lay claim to logic and incisiveness as particular attributes - none of your 'airy fairy' nonsense in our discipline. These are worthy qualities that in practice are too often distorted by political or self-interested motives. Jane would have delighted in exposing such distorted behaviour to ridicule. She is celebrated for the nicety of her language and, preferring the rapier to the bludgeon, she could use it in a most cutting manner - there is plenty of evidence for this in her private correspondence. With Jane, there is no self-indulgent ornamentation beyond the immediate purpose of her stories; she would have wielded Occam's razor with admirable zeal.

All very fine and pretty you may say, but could not the same be said of other celebrated authors? Dickens or Shakespeare, say? I maintain not. Part of the appeal of engineering lies in the appreciation of form (whether physical or virtual) and function. There is a satisfaction to be found in identifying solutions, which are fit for purpose whilst employing an economy of resource. There is also the appreciation of the harmony of forces or processes that inform our designs and empower our solutions. There is poetry here that is an emergent property of our structures and systems. In Jane's stories there is remarkably little of the directly poetic; although their subject may be romance, there is nothing romantic in their composition. She established free standing structures with a discipline and focus which yields poetry as an emergent, rather than an intrinsic property. Like an engineer.

# 16
# BEYOND ALARP

Under certain circumstances, the Health and Safety Executive will look for what they consider to be appropriate safety provisions regardless of whether a risk assessment identifies them as beyond ALARP (As Low As Reasonably Practicable). This may seem an extraordinarily high-handed stance, but if we look a little more closely at the circumstances we can see it is not an unreasonable position.

It is typically when the scenario under consideration represents a high hazard potential AND the associated risk assessment has high uncertainty, that the HSE will invoke this philosophy. Sensitivity analysis can be usefully employed to examine the impact of underlying assumptions or estimates on risk assessments ('if I wiggle this input, how much does the output move?'), but this may not always resolve matters sufficiently.

This is potentially fraught territory of course; a duty holder might consider that the regulator is being unduly conservative and these differences in opinion cannot be resolved by resorting to calculation; by their very nature they require judgement to be exercised. It may be that the regulator does not immediately know that this is the circumstance; it may be that emerging misgivings cause him to call for increasingly conservative assessments that, to the duty holder, appear to be increasingly divorced from reality. You can see the scope for difficulties here.

Certainly, conservatism is understood to be prudent in matters of safety; fair enough. But this can be overplayed. If conservatism is invoked at multiple points in an assessment, the compounded effect can be a massive distortion in the findings. And of course, conservatism doesn't cost the regulator anything. Attempts by the duty holder to argue against conservatism in order to contain spiralling costs may be characterised as 'economising on safety', but this is mere spin. Rather, if we consider that the expenditure and resources might be misplaced and would produce a better safety return if spent in other ways, we can gain a different perspective. That said, if the argument is for/against some provision that would be normally be considered standard good practice, the regulator's default position will, understandably, be for adoption.

Clearly, the regulator has the whip hand here, and, typically, through continual exposure to risk issues, he might be expected to be in the better informed position and well placed to educate duty holders. This is not universally true however, and the specific circumstances might not conform to established precedents and duty holders might well have an alternative view that is nevertheless considered and responsible. The regulator should be willing to engage in a dialogue to explore the possibilities with something of an open mind. I imagine most duty holders are concerned to do the right thing, and if they should fall short in this it will be through ignorance rather than a wilful disregard. We can have no sympathy with those that are 'trying to get away with it'. Particular difficulties might arise if egos become invested in pronouncements or, if questioning approaches even if offered in good faith, are seen as impertinent challenges to authority. Positions may become entrenched and dialogue supressed. The key point is that there should be a considered, informed dialogue between mutually respectful professionals.

# 17
# INTERNET FORUMS FOR PROFESSIONAL GROUPS: USES AND ABUSES

Internet forums offer a potential resource for continuing professional development, but they are not without their hazards. I would characterise them not as reference sources, but rather as exploration tools.

I met an example recently where an engineer, who identified his role as 'senior' (therefore denying himself the excuse of novicehood) posted an enquiry to a functional safety forum with the question 'How can I determine test intervals for safety instrumented systems in order to meet the required safety integrity levels?'. This is the functional safety equivalent of asking 'How do I work out the current in a resistor if I know the voltage across it?'. I find this troubling: I believe any professional engineer should conduct primary research through recognised references or publications, or through consultation with some suitable authority (i.e. through diligent homework), not through some lazy, off-hand enquiry that is broadcast to the world.

The worry is that some ill-considered or misplaced response might be accepted as definitive; any engineering works based on the new 'understanding' could be correspondingly flawed. Any engineer that thinks a broadcast enquiry through a social network is a way to acquire underpinning knowledge and understanding is just the person to blithely press on without validation of their 'learning'.

If an enquiry relates to some fundamental point there is the danger that it will be beneath the interest of most forum members and that the first half-baked reply will remain unchallenged. International contributions may be from people who are not writing in their native language, with all the resulting opportunities for something to be lost in translation. Even if the contributions are in the same native language, there remains broad scope for misunderstandings. Respondents may answer the question they thought you asked, the question they think you should have asked, or the question they prefer to answer. And, of course, the original posting may itself be ambiguous or otherwise confused.

Postings are often 'banged out' on an off-hand basis without proper consideration or without observation of the usual professional courtesies. Perhaps it is the immediacy of the format that invites this (an invitation that is enthusiastically accepted by many), but as professionals we are supposed to know and behave better. In any professional forum, whether online or not, we should adopt appropriate professional disciplines: cyberspace should not be thought of as some kind of playground where sloppiness or rudeness is acceptable. There is an important distinction to be drawn here between informality and sloppiness; professionalism and informality are not mutually exclusive; professionalism and sloppiness are. "Hi" or "You must be joking!" may be acceptable (although we should not presume an inappropriate familiarity); "The square on the hippopotamus is sometimes equal to the sum of the rectangles on the other two sides" is not, unless explicitly offered as tentative speculation.

In recognising the potential hazards, the first rule must be *'caveat lector'* (let the reader beware). I would rank discussion in a forum on a par with that in a bar; there is likely to be the same range of sense, nonsense, half-truth, invention, wishful thinking and delusion. There is a corresponding requirement to evaluate each contribution: Does it make sense? Is it from an authoritative (and sober) source? Is it from some maverick who likes to 'mix it'? We should be particularly wary of contributions made under nicknames; any forum related to professional concerns should be conducted on the basis of real names, and affiliations, where pertinent, should be declared. (A case could be made for making this a wider rule – if you are not prepared to put your name to a contribution on a forum, perhaps you should not be posting at all. Some people believe that anonymity provides a licence to commit any number of abuses. We can see parallels with the unrestrained behaviour of the anonymous individuals that constitute a mob; a breakdown of civilisation.)

Consultation with one's peers is an entirely legitimate and, indeed, necessary activity; it provides a means of calibrating our judgements and validating our thinking, and internet forums clearly have a potential part to play in our continuing professional development, so the questions arise:

a) How should a response be constituted? - If our response is speculative this should be made clear. If we make assumptions, they should be explicitly declared. If our response is valid for a restricted domain, we should identify its boundaries.

b) How should a response be gauged? - Any response received needs to be critically assessed for coherence; is it self-consistent and consistent with existing learning? Any critical information provided by a respondent should be validated against independent references.

c) How should an enquiry be formulated? - Enquiries should, of course, be unambiguous, and should identify the context if that is material. Don't use a forum to ask what 2+2 equals: use a forum to find the pros and cons of calculating the sum with a calculator/with a spreadsheet/with an application program.

The value of these forums is typically not in providing definitive answers, but rather in providing new considerations, new lines of enquiry, new possible approaches, new places to look. They provide starting points, not end points.

# 18
# LOGICAL V SENSIBLE

I recently encountered a good example of the sort of nonsense that can arise from the indiscriminate, uncritical reference to standards.

In matters of explosive atmospheres, there is a requirement to perform an initial detailed inspection of new installations of 'Ex' protected equipment, notionally before they are first put into service. Fair enough.

A client of mine was advised by an 'expert' that he should retrospectively (25+ years later!) perform a detailed inspection of his 'Ex' assets. Now this advice is logically consistent with the stipulation in the relevant standard (IEC 60079-17), but logical is not synonymous with sensible. My client manages a legacy plant with something north of 5000 assets; none of which have a record of any initial inspection. They are, however, subject to a periodic, non-invasive inspection, together with a sample detailed inspection, and experience has shown that there is no reason to suspect any deficiency.

The notion that these 5000+ assets should now be subject to a detailed inspection is absurd. It would mean a massive programme of invasive disturbance to the installation, which would most likely result in net harm; enclosures could be compromised, wiring and connections damaged. Many of the provisions for 'initial inspection' would in truth be redundant in that the associated defects they are intended to identify would have been revealed by failure in service.

In a similar vein, another 'expert' advised a client that if they undertook any work in a cubicle or junction box that contained any safety loops, then every such loop must be tested after the work is completed – regardless of the failsafe provisions in their design, or whether they have been subject to any interference.

But here is the problem: this advice is being formally tendered by an 'expert'. The logic deployed may be irrefutable, which, coupled with the supposed authority with which it is stated, makes it appear entirely plausible, or even worse; unquestionable.

Time and again I find myself defending my clients from just this sort of nonsense; they have been told by a supposed authority that they MUST do such and such. Closer examination of the circumstances and the cited references shows that there is no 'must' about it. It is the role of the professional engineer to evaluate such assertions and distinguish between 'logical' and 'sensible'.

'Logical' simply means consistent with the premises; it does not mean 'true'!

The syllogism:

*all things with four legs are animals*
*this table has four legs*
*therefore this table is an animal*

is entirely logical. Hopefully we can agree it is not sensible!

The premise 'all things with four legs are animals' is evidently false, but the falsehoods we meet in our work are usually more subtle. More typically a premise is asserted which, on a standalone basis, looks reasonable, but which is made without due consideration of the context in which it is made.

Be warned. Just because a proposition is logical does not mean it is sensible! Many 'experts' appear to operate in a kind of vacuum that insulates them from the practical implications of their advice.

# 19
# MATHS FOR MATHS' SAKE?

Mathematical rigour is a good thing. What is your take on that assertion? At first glance it appears uncontroversial, but with further thought we discover that rigour does not sit alongside motherhood and apple pie as a universal good. If it obfuscates or demands additional resource without providing material benefit, it is not earning its keep. As with most things in engineering, context is everything. (This could even be regarded as the key difference between engineering and physics; in physics the context is…the universe. In engineering we usually work within a more restricted domain.)

This philosophical concern cropped up in a debate on an internet forum regarding functional safety (a large part of my 'day job'). Mathematical models are routinely used to calculate the availability of protection systems, and often some shortfall in mathematical rigour will be challenged and furiously debated. Now, if mathematical rigour 'floats your boat', there is a world of fun to be had. There are some intriguing aspects to these debates, but these are usually 'a sideshow to a sideshow' (to borrow Lawrence of Arabia's description of his First World War exploits). People are debating mathematical models here; by definition, approximations to reality. There will always be some shortfall in the outcome regardless of the degree of rigour employed, in the same way that any map with a scale of less than 1:1 will be missing some detail.

I don't say these debates are meaningless or without interest or value, but they are often irrelevant in the context of the primary mission. As engineers we should be wary of the pursuit of rigour for its own sake, or of those that would urge this upon us. I'm reminded of a conversation about a nuclear fusion reaction in which it was reported that it took place at a gazillion degrees, 'but I can't remember whether that's Celsius or Kelvin'(!).

As engineers, our primary concern is fitness-for-purpose rather than some abstract rigorous 'truth'. If it works, that is truth enough. As technical professionals, the abstract ideal may well intrigue us, but if we allow ourselves to be seduced by the siren calls of rigour and the pursuit of 'ultimate truth' we will have betrayed our core purpose. The models are a means to an end, not an end in themselves (not for engineers anyway).

The true transcendental beauty of mathematics is most evident in its pure abstract form. Our rough manhandling of it in our utilitarian fashion might be offensive to some, but boy, just look at what we can do with it! There is a readily apparent beauty in a suspension bridge or an aircraft that is underpinned by the mathematical principles their designs exploit (used and abused as they are by our civil and aerospace cousins). However, although less apparent, that is no less true of the thermodynamic, chemical and fluid mechanical 'ballet' performed in our processes.

Do I wax too lyrical here? Possibly, but a search for an optimal engineering solution is often informed by an appreciation of elegance; an elegance that derives from the underlying mathematics.

# 20
# PROFESSIONAL COMPETENCE

Continuing Professional Development (CPD) is a critical concern for all professional engineers, whether pursuing registration, already registered, or acting in a professional capacity without registration. Given the thrust of many CPD schemes as currently promulgated by the engineering institutions, you could be forgiven for assuming it is only a matter for aspiring registrants; you would, however, be wrong. Certainly those pursuing registration must demonstrate Initial Professional Development (IPD), but they must also demonstrate an ongoing (post-registration) commitment to CPD. Established registrants must also have an ongoing commitment to CPD, although for the mature engineer it would be less a question of establishing competence and more one of maintaining it. Without this commitment there is the real danger of increasing obsolescence and a loss of competence. If you do not effectively engage with CPD you can have no real claim on the title 'professional engineer'.

## Code of Conduct

The Engineering Council in the UK has published a CPD code of conduct for registrants which says:

> *Engineering Technicians, ICT Technicians, Incorporated Engineers and Chartered Engineers should take all necessary steps to maintain and enhance their competence through continuing professional development (CPD). In*

*particular they should:*

*1. Take ownership of their learning and development needs, and develop a plan to indicate how they might meet these, in discussion with their employer, as appropriate.*
*2. Undertake a variety of development activities, both in accordance with this plan and in response to other opportunities which may arise.*
*3. Record their CPD activities.*
*4. Reflect upon what they have learned or achieved through their CPD activities and record these reflections.*
*5. Evaluate their CPD activities against any objectives which they have set and record this evaluation.*
*6. Review their learning and development plan regularly following reflection and assessment of future needs.*
*7. Support the learning and development of others through activities such as mentoring, and sharing professional expertise and knowledge.*

This may appear to be a demanding commitment, but take heart, the task need not be so very daunting; it is likely that you are continually engaging in CPD – you may simply fail to recognise (and record) it. If at any time, in support of an assignment, you undertake some research (e.g. into a standard), or consult with your peers (e.g., in examining proposed approaches), you are engaged in CPD.

CPD does not have to be by so called 'structured' means (e.g. training courses); self-directed study and incidental learning may well be the predominant means of undertaking CPD. One of the key points is the 'reflective' aspect referred to in the Engineering Council Code; why this emphasis on reflection? I have yet to find a better explanation than Steven Pinker's when writing (about writing) in his book *The Sense of Style*: 'As with any form of mental self-improvement, you must learn to turn your gaze inward, concentrate on processes that usually run automatically, and try to wrest control of them so that you can apply them more mindfully.'

Reflection in the context of CPD is the considered review of any learning acquired (by whatever means). This allows the learning to be consolidated into your wider framework of knowledge, understanding and competence. It also allows consideration of possible directions for further development and identification of corresponding actions. This is where the practice of recording CPD is helpful; it facilitates the reflection on the learning outcomes. It also allows a ready means of substantiating your commitment to CPD.

## Recording CPD

Compiling a record of CPD undertaken is key component of compliance with the code of conduct, but is typically the aspect where many engineers will fall short. CPD activity may be categorised in many ways; the following basis might be helpful:

- Structured: e.g. training courses, conferences, workshops, lectures.
- Incidental: acquired incidentally through the prosecution of work assignments (the specific scope should be identified).
- Self-directed: e.g. independent individual study - books/papers/articles/standards. Consultation on specific issues with peers. Development of papers/articles/presentations.
- Professional Engagement: involvement with the wider profession and professional practice, typically other than through work e.g. through professional bodies, committee work, standards panels. Also event attendance e.g. section meetings, conferences or workshops. Such engagement allows validation of your K&U through interaction with other professionals beyond the immediate workplace.

A record of development outcomes might be categorised as follows:

- New skill.
- New knowledge & understanding (K&U).
- Consolidation: placing existing K&U into a more complete and coherent framework.
- Validation: Confirmation or calibration of existing K&U (essentially by cross checks with independent professionals).
- Awareness: knowledge of existence of techniques/considerations/scope without specific

competence.
- Qualification: Certified development typically from formal assessment following training.

## Measuring CPD

There is considerable debate about the means of measuring CPD; the ideal would be to measure 'outcomes', but these are of such disparate natures and involve such broad uncertainties that a meaningful aggregate measurement presents serious difficulties. The measurement of 'inputs' is a more practicable proposition, but these 'inputs' are only indirectly linked with the 'outcomes'. That said, given the establishment of underpinning knowledge & understanding (UK&U) that is one of the foundations of registration, we might anticipate that competency outcomes will correlate well with the investment in inputs. It is a relatively simple matter to record hours invested in individual development topics which may be categorised as identified above. Outcomes might be identified on the basis of the achieved level of competence.

## Professional Practice

The specific technical competencies needed to effectively fulfil any given technical professional role will be largely self-evident. But there is more to being a professional engineer than the accretion of a certain technical skill set; there is also the proper appreciation of the wider context in which this skill set is to be exercised. This we might identify as 'professional practice', i.e. the prevailing expectations, professional & ethical standards, working practices and obligations, that any engineer worthy of the name 'professional' will need to conform with. This should form a default development field or topic against which CPD activity should be recorded for any professional engineer, regardless of their specific professional role. This field may perhaps be most readily addressed by the 'professional engagement' category of development activity as described above.

Congratulations if you have made it to this point - you have just completed some CPD; time now to reflect.

# 21
# CROSSING THE LINE?

Prosecution of individual engineers (as distinct from corporations) is very rare. Typically, these cases arise from instances of gross negligence; a wilful disregard of health and safety concerns. In truth, anyone showing such wilful disregard can hardly be considered 'professional'. I have not researched this particular point, but I imagine that usually those individuals found guilty have taken it upon themselves to perform the role of a professional engineer, but without the professionalism or competence.

No one, in the course of legitimate business, sets out to do harm. The harm is an inadvertent consequence of decisions we make (or don't make) or actions we take (or don't take). There are sins of both commission and omission.

There can't be many engineers involved with potentially hazardous circumstances that have not relaxed their vigilance or diligence at some point, typically when persuaded that the likelihood of harm is remote and intervention would be burdensome or irksome. A very few will come to rue their decisions when the Swiss cheese holes rotate into alignment. It would be naïve to universally criticise this behaviour; there is a judgement call to be made. Insistence on an absolutely compliant (by the book) approach may under certain circumstances bring its own hazards, or may, if the measures are seen as disproportionate, undermine the broader safety culture within an enterprise and actually promote the taking of short cuts. It might undermine the credibility of the insisting engineer, whose future ability to influence things to the good may be diminished (a subtle form of crying 'wolf!').

The challenge is to recognise when a line is about to be crossed. The line will not always be well defined, but is usually recognisable (although sometimes only in retrospect with 20/20 hindsight). The line is not fixed either; it will move with circumstance – the degree of risk in particular.

Given the nature of our industry, it is perhaps no surprise that it is a struggle to find prosecutions of individual engineers, but let me illustrate the question with reference to the tragedy in 2004 at Tebay (UK), where four railway workers were killed and five injured when a runaway trailer ran down an incline in the dark. The hydraulic brakes on the trailer had been disabled and in lieu of these the trailer wheels had been chocked with timber; subsequent operations disturbed the trailer sufficiently to dislodge the chocks. Two men were subsequently found guilty of manslaughter, essentially, I understand, for wilful disregard of health and safety regulations.

When was the line crossed here? I do not know the codes of practice or standards that pertain to such affairs. I can imagine myself seeing the chocks in place and perhaps thinking 'that trailer is not going anywhere'. (Just to be clear I haven't seen the chocks and have no idea of the actual arrangement used.) This immediately points to questions of competence – engineer I may be, but I am not competent in railway standards and practices.

Was it when the brakes were disabled that the line was crossed? Presumably if some other high integrity arrangements had been made that would have been acceptable? I might have hesitated over the chocks, but some suitable stop, clamped and locked to the line? Perhaps this is the rail equivalent of isolation through a double block & bleed or a blank? Possibly the chocks were expedient in allowing quick removal for repositioning of the trailer? The equivalent of a closed hand valve.

It was said that commercial and time pressures were contributory factors. Reading the reports of the incident, it seems that the trailer's hydraulic brake lines had been filled with ball bearings to make it appear that all was 'above board'; I don't understand it either (perhaps to simulate incompressible fluid?), but such deliberate attempts at deception are unquestionably over the line.

It is certain the men did not set out to harm anyone; no one ever does. Of those that do cause harm, we might ask were they:

- Not competent and unlucky? (In which case whoever authorised them starts to look like they may have a degree of culpability.)
- Competent but VERY unlucky? (We can sympathise.)
- Notionally competent and only a little unlucky? (A damning combination.)
- Notionally competent but with a wilful disregard for safety? (Bang to rights.)

Ah! For 20/20 foresight.

# 22
# TO SHRUG OR NOT TO SHRUG?

Please forgive the vulgarity, but I don't think there is any denying the prevalence of the slang phrase, 's\*\*\* happens'. (Don't be misled by my title; it is not one of Shakespeare's, and I agree it lacks any redeeming wit, but it does seem to have been given popular impetus by the 1994 film *Forrest Gump*.)

S most certainly does H. It may be true, but is there any wisdom here? The phrase is typically invoked in acknowledgement of, or encouragement of, resignation and acceptance; there is 'nothing to be done about it'. The implication being that we should shrug and move on. There are circumstances where this may well be an appropriate response, but there is the danger that this attitude may be invoked inappropriately, when there is in fact 'very much something to be done about it'.

In our professional lives, if we see critical systems or procedures being degraded or compromised we cannot 'shrug and move on'; or at least not if we wish to lay claim to the title 'professional'. Blatant action in contravention of authorised policy or procedure may be immediately apparent and readily challenged, but what of inaction? Sins are more usually of omission than commission. The temptation is to turn a blind eye or 'walk by on the other side'. Again, there may be circumstances when this is an appropriate response, but we should be persuaded that this is for some greater good rather than the avoidance of personal discomfiture.

The more insidious threats arise from the slow, incremental degradation of our operations. Any individual decrement may not be seen as that significant (if perceived at all); hence the insidious nature of the threat. It is the cumulative effect of these changes that is the real concern. In terms of process safety, this corresponds with a progressive widening of the holes in our Swiss cheese model. Continual vigilance is required. Note that these changes are not confined to physical plant and equipment; they may well be in practices and procedures – an increasing acceptance of sloppiness.

Why fight it? (S*** happens!) We fight it because we have a professional duty to do so, and in recognition that this is not some inevitable mechanism that we are powerless to influence. Anytime we hear someone (or ourselves) invoking this philosophy, either literally or metaphorically, we need to do a quick sanity check – is there really 'nothing to be done' or are we ducking our responsibility or condoning abdication by others? Only if we are truly satisfied that there is nothing to be done may we smile/grimace and shrug.

# 23
# THE DUMB CUSTOMER

The UK's Health and Safety Executive point to the need for duty holders to act as 'intelligent customers' when engaging others to undertake work which may have implications for health and safety. The philosophy can, however, be usefully extended to other areas.

We may approach the characteristics of the intelligent customer by examining those of the 'dumb customer', who will:

- Abdicate responsibility.
- Assume the contractor knows what is needed.
- Assume the contractor knows what he's doing.
- Assume the contractor has done/will do it.
- Retain no record of what has been done.

One of the primary motivations of the 'dumb customer' is a misplaced, (and ultimately futile) desire to transfer responsibility/liability to another party. A key point is that if failures or deficiencies do arise, the blame cannot always be (or not in the eyes of the law anyway) laid at the door of the contractor. A failure does not necessarily mean that the contractor has been incompetent, unscrupulous or exploitative. A contractor may, by his own lights, be acting in good faith, in accordance with the requirements as he understands them.

## The Dumb Customer

From these considerations we may go on to identify the characteristics of the intelligent customer:

- He understands and specifies what is needed.
- He exercises due diligence:
    - He satisfies himself of contractor competence.
    - He maintains proper overview (supervises and reviews).
    - He co-ordinates activities (ensuring nothing falls between the cracks).

One common manoeuvre is to issue a list of standards that the contractor/vendor 'must comply with', that is as long as your arm. Such lists are often used without real discrimination and sometimes with mutually exclusive provisions! The thinking appears to be that this will form some sort of insurance policy; a stick to beat the (wily rogue) contractor with. Perhaps it will, but it does seem a mean spirited basis on which to enter into a contract. If you are persuaded you need this stick, perhaps you are issuing your enquiries to the wrong contractors?

The requirement for an 'intelligent contractor' (I believe I coin the phrase) is not explicitly made, but such an animal will provide matching support in recognition of the intelligent customer's responsibilities. What might this support look like? I suggest:

- Intelligent consultation over the customer's needs (rather than uncritical acceptance of a procurement specification).
- Identification of key interfaces; what interactions will be necessary for successful delivery?
- Identification of required competencies (what skill sets are needed?) and maintenance of a competence register.
- Identification of requirements of deliverables in support of the intelligent customer's responsibilities (what will the customer need post contract close out?) with full version control and traceability.

Nothing so very profound in any of this, but without a specific focus on these concerns there is a possibility that inadvertent gaps will develop.

We might identify two further key attributes of the intelligent customer; a willingness to pay a premium for an intelligent (rather than dumb) contractor, and a recognition that the contract should be collaborative rather than competitive. If the contract is approached as a zero sum game (like chess; what one player gains, the other loses) this will militate against a constructive partnership. A win-win is entirely possible and much to be preferred. Unfortunately, there are those players in the 'game' for whom the only way to be confident of a 'win' is to make sure the other guy loses, and who make it their mission to 'screw the contractor'. With this mind-set however, the much more likely outcome is lose-lose. A very unintelligent approach.

# 24
# ENGINEERING JUDGEMENT SPACE: FOUR SHADES OF GREY?

The effective exercising of engineering judgement may be identified as one of the hallmarks of the true professional engineer. But in embracing this notion, there is perhaps a danger that individual engineers may become over eager to exercise their judgement in demonstration of their professionalism. We may identify a further hallmark as an awareness of this danger and the recognition of the need to guard against it. The all too human traps of arrogance and hubris are set to ensnare us. (We should do whatever we can to prevent the marriage of expertise with arrogance; their offspring are doomed to be most unbecoming!). We need to exercise appropriate meta-judgement; judgement about judgements.

Our primary defence must be an awareness of this situation, and in this regard, it may be useful to consider the 'space' within which engineering judgement is exercised by the individual engineer. We may ask what dimensions characterise this space? I propose two: sensitivity and domain familiarity.

1. Sensitivity; how sensitive is consequence to an error in judgement. How severe would the loss be? To what degree would the objectives be compromised?

2. Domain familiarity; how novel or familiar is the context within which judgement is to be exercised? Is the terrain well understood and

well-trodden?

In Chapter 12, judgement error was identified (using the differential form) as being due to inaccuracy in the engineering model employed and the data used:

$$\delta judgement \approx \frac{\partial judgement}{\partial data}.\delta data + \frac{\partial judgement}{\partial model\ inaccuracy}.\delta model\ inaccuracy$$

If we characterise sensitivity as

$$\frac{\partial consequence}{\partial judgement}$$

we may extend this representation:

$$\delta consequence \approx \frac{\partial consequence}{\partial judgement}.\left\{\frac{\partial judgement}{\partial data}.\delta data + \frac{\partial judgement}{\partial model\ inaccuracy}.\delta model\ inaccuracy\right\}$$

The underlying assumption here is that domain familiarity leads to a more accurate model and/or better data. Note that the model may not be an explicitly mathematical one, it might very well be 'fuzzy' and intuitive, but the expectation is that it captures the pertinent relationships.

Given these dimensions we may chart judgement-space with a range of validation measures:

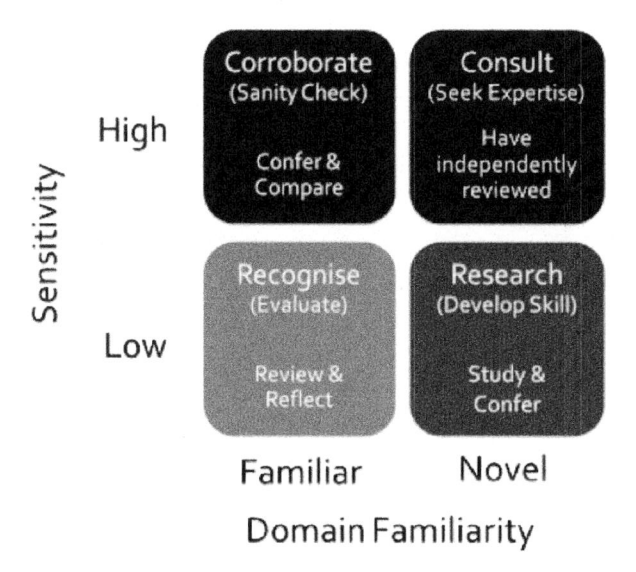

Engineering Judgement Space: Four Shades of Grey?

|  | Familiar | Novel |
|---|---|---|
| **High Sensitivity** | Corroborate (Sanity Check) — Confer & Compare | Consult (Seek Expertise) — Have independently reviewed |
| **Low Sensitivity** | Recognise (Evaluate) — Review & Reflect | Research (Develop Skill) — Study & Confer |

Domain Familiarity

There is of course a continuum of grey here, but in characterising the space with discrete zones we help make the space more immediately accessible (It was tempting to go with a 7 x7 matrix, giving very nearly 50 shades of grey, but common sense prevailed):

- If the space is well-known to us, and the sensitivity to any error would be low, the individual might rely on him/herself to validate their own judgement, but it would still be appropriate to recognise that judgement is being exercised and to make an effort to consciously evaluate this through review and reflection.

- If the space is familiar, but the potential sensitivity is high, then it behoves us to corroborate our judgements with some form of sanity check to reveal any inadvertent blind spot and confirm that the space does not have has some unfamiliar characteristics (hidden valleys perhaps?) This might typically be through some form of peer-review with colleagues and comparison with previous cases.

- If the space is novel and therefore not familiar to us, but the sensitivity to an error would be low, then we might undertake to research the space with appropriate continuing professional development (CPD) activity, including peer-review with colleagues. The space may become known to us, in the sense that we have studied the 'terrain', but it would not become 'well-trodden'; our experience would remain limited.

- If the space is novel and of high sensitivity, then formal consultation with independent expertise might well be appropriate.

It might be argued that instead of domain familiarity, 'confidence' might have been adopted, but confidence does not necessarily correlate with ability. In his book *Thinking Fast and Slow*, Daniel Kahneman also discusses the nature of expert intuition and reports that the factors that point to a reliable intuitive judgement from an expert are:

A. 'an environment that is sufficiently regular to be predictable', and
B. 'an opportunity to learn these regularities through prolonged practice'.

He makes the key point that the confidence of the person proclaiming a judgement is a very unreliable guide to validity and should be disregarded. We should consider rather whether the field they are pronouncing upon is regular and whether they have the necessary experience. It is the combination of a regular 'environment' and 'prolonged practice' that constitutes 'domain familiarity'.

There are a variety of cognitive biases that may influence our judgement and it is important that we should be aware of these and consciously strive to counter them. (I urge that this should be a formal part of the curriculum for engineering students.) Our judgements should be based on a proper understanding of the relevant causal mechanisms and we should avoid confusing correlation with causality.

There remains then the question of calibration; what constitutes high sensitivity or a novel domain? This is resolved through broader engagement within the profession; collaboration with colleagues, peer-review, the technical press, conferences, exhibitions, standards, professional committees, professional institutions etc. In a 'word' (acronym); CPD.

In closing, let me say that I recognise that the map is susceptible to parody:

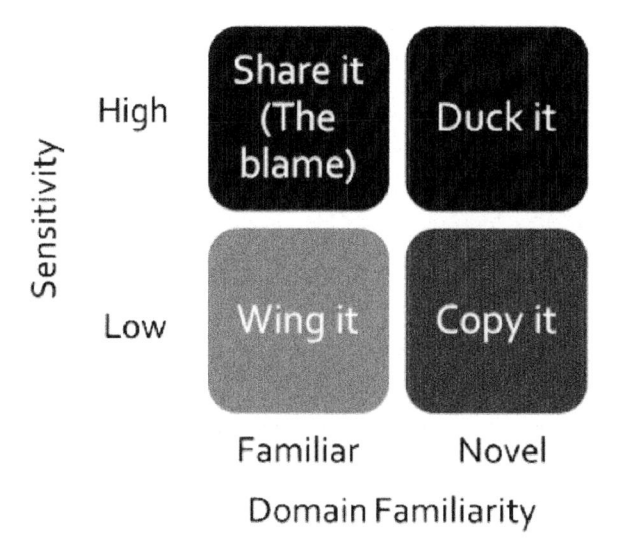

|  | Familiar | Novel |
|---|---|---|
| **High** Sensitivity | Share it (The blame) | Duck it |
| **Low** | Wing it | Copy it |

Domain Familiarity

# 25
# ENGINEERING A JOKE?

In 2015, a scandal broke in which it emerged that Volkswagen had engineered an emissions testing 'defeat' in which some of their diesel engine cars were programmed to recognise when they were being tested and modify their performance to reduce emissions. It would be interesting to understand how this scandal came to be. You can well imagine a jocular conversation in a bar, something along the lines of, 'Hey! I know what we could do…'. A conversation born of a heady mix of beer, exasperation with the perversity of the standards, and the unreasonable demands of 'management'. Nothing so remarkable there.

The really interesting point is this: how did a drinking jest turn into an engineered 'solution'? Clearly, the joke that turned out not to be a joke, must have been endorsed by a team and sanctioned by management. It could not have been the work of a lone wolf. But apparently no one said "Whoa now! - You must be joking!" I can only think that the standards and testing regime had become so divorced from the realities of actual performance (one hears stories of stripped down and specially finished cars being submitted for testing), that the 'solution' no longer appeared to represent an unacceptable crossing of a line. After all, the results were real, they were not being falsified, they represented the actual capability of the engines; it was just that the configuration did not match that deployed in normal use.

When the 'chickens come home to roost', the classical approach would be for management to absolve themselves of blame and for the engineers concerned to be identified as rogue elements, but to be fair to VW, they have acknowledged corporate failings.

This territory represents perhaps one of the most severe tests of professionalism; the pressure from management; 'I'm not interested in your excuses. This project has to be delivered by___. The company's future/workforce security/your prospects are resting on this'. The pressure will be intensified if any delays or overruns lie within your area of responsibility, regardless of whether there was anything that you might prudently have done differently. In our own defence we should not allow unrealistic expectations and delivery programmes to remain unchallenged and should insist on recognition of uncertainties in project delivery.

In the face of such pressures however, a degree of compromise might well be reasonable. But when should you refuse to take a short cut or lower standards? There comes a point where the only honourable course would be to tender your resignation, or refuse cooperation and invoke disciplinary action, but where on the continuum of professional behaviours does this point lie?

The simple test is that of hypothetical public scrutiny; if a full account appeared, how would a reasonable audience regard your behaviour, given that they were suitably informed of the context and had an appropriate degree of understanding? I must stress the 'hypothetical' here; there is little prospect of such a balanced assessment, but we cannot allow the usual media distortions to be a factor in our evaluation (that is very hazardous terrain that, mercifully, our profession requires us to leave the politicians to negotiate).

It is only a sense of honour (and its converse – shame) that can sustain us through such trials of our professionalism. It seems that popular recognition of these values is in serious decline, as the talentless and shameless are increasingly 'celebrated' for their very shamelessness. But we are not appearing on a 'structured reality' TV show. We are not trying to 'keep up with the Kardashians'. We are endeavouring to make things work in 'real reality'. That brings with it certain professional responsibilities – these are truly worth celebrating and we should embrace them and hold them dear.

# CODA:

# TO BOLDLY ENGINEER...

Engineers do not seem to be celebrated in film the way that lawyers and medics are. Presumably because these professions are associated with circumstances that most people can relate to in some way, even if by 'imagined' rather than direct experience.

Engineers do find some limelight in the sci-fi arena, but even then it seems typically to be in a peripheral way. 'Scientists' (although a relatively new invention, the term first being coined in the 19th Century) appear to be a more popular fictional representation of the technical professional than 'engineers' (a much older term from the Latin *ingeniare* 'to contrive or devise' and *ingenium* 'cleverness'). Perhaps the engineer character that most immediately springs to mind is Scotty from *Star Trek* - the dour Scotsman who worries about whether his engines can 'take any more' and the condition of his dilithium crystals. (There is a proud tradition of engineering within Scotland, but no actual experience with warp drives that I am aware of. However, in passing note that dilithium is a real molecule formed from the bonding of two lithium atoms.)

Interestingly, although Scotty, as chief engineer, was a much loved and celebrated character in the original series from the 1960s, when the franchise was re-launched with *Star Trek: The Next Generation* in the 1980s, there was initially no place for the chief engineer as a main character; instead we had an empathic 'counsellor' in the shape of

Deanna Troi. It seems however, that the screenwriters realised that fictional engineering matters were key to plot development and subsequently the character Geordi La Forge, originally a helmsman, was migrated into the post of chief engineer. The substitution of a counsellor for an engineer was perhaps a sign of the times? In the 1990s series *Star Trek: Voyager* it seems the lesson had been learned; after the pilot, the following episode revolves around who should be assigned the post of chief engineer. B'Elanna Torres is the eventual choice - a coup for the promotion of women in engineering (or half-Klingon women at least). She remains a central character throughout the series.

One character that is unquestionably an engineer first and foremost is that of Tony Stark: *Iron Man*. Originally a Marvel comic book character, he has found a new lease of life with a series of 'blockbuster' films.

From the UK, there is the long-running TV drama *Doctor Who*. The Doctor is a quirky but nonetheless heroic character with unmatched intelligence and technical prowess. He is not explicitly referred to as an engineer but the clues are all there; he is continually reconfiguring systems in order to save the universe and his most intimate relationship is with his sentient time travelling spaceship, the TARDIS (an acronym for Time And Relative Dimension In Space), which is an artefact of 'transdimensional engineering' that is 'bigger on the inside'. If you want further proof, he even wields a sonic screwdriver! He captivated my imagination as a child (closely followed by the logical Mr Spock) and I retain affection for him to this day. (I proudly wear the *Dr Who* socks my sister Sally gave me for Christmas.) I am delighted to claim him for our profession.

# APPENDIX:

# TWENTY THINGS TO REMEMBER

1. Professional engineers do not guess, they estimate. Any 'engineering judgement' made without a rationale is a guess. (Ch.1)
2. If a proposition does not make sense, either it is flawed or you have not understood, find out which.
3. Have the courage to ask 'stupid' questions that may well turn out to be not so stupid after all. (Ch.12)
4. If analysis and expectations do not match, refine one or both until they do. (Ch.2)
5. In matters of risk, calculation should inform judgement, not replace it. (Ch.2)
6. Never underestimate the value of a sanity check; we all suffer blind spots. (Ch.1)
7. If you wish to be regarded as a professional, act like one. (Ch.4)
8. Take some trouble with your communications. You know what you mean; make sure the other guy does too. Write your memo so that he can't misread it, even if he wanted to. (Ch.4)
9. Always consider the sensitivity of your evaluations to inaccuracies in both data and model. (Ch.12)
10. Only give up CPD if you wish to give up engineering. (Ch.10)
11. Regard your professional integrity as a gift to yourself; you cannot be robbed of it, you can only choose to give it away. (Ch.4)
12. Never use hope as a design tool. (Ch.9)

13. Do not confuse experience or qualifications with competence. (Ch.10)
14. Don't take your business hat off when you put your engineering hat on. (Ch.14)
15. Don't take your engineering hat off when you put your business hat on. (Ch.1)
16. Stay aligned. Always, always consider 'what are we trying to achieve?' (And why?) (Ch.14)
17. Logical does not necessarily mean sensible. Be wary of 'clause quoters'. (Ch.6 & Ch.18)
18. We are engaged in engineering, not physics; anything beyond the third significant figure (except when dimensioning), is usually a joke. (Ch.5)
19. Sometimes a problem can be usefully considered from a new perspective - a bar stool for example. Mere flippancy? Well no, some such manoeuvre may allow your thinking to jump from a well-worn groove.
20. Explaining an issue to someone else will oblige you to explain it properly to yourself. This can be surprisingly effective in providing new realisations. Try it in a bar.

# ABOUT THE AUTHOR

Harvey T. Dearden BSc CEng FIET FInstMC FIMechE FIChemE has worked in the process industry sector for over 35 years and has been employed by a variety of vendors, contractors, consultants and end users. He now manages his own consulting practice, *Time Domain Solutions Ltd.* (www.tdsl.org.uk), providing support in matters of process control and protection, and associated risk assessment and management concerns. He is known to many through his involvement with professional engineering institutions and his numerous papers and articles. Previous works include *Functional Safety In Practice* (2016) and *Crowns & Coronets, Mitres & Manes* (2016). He is married to a Welsh (Anglesey) girl who prefers not to be carried over the border and so he continues to live in North Wales, where there is hardly any process industry (!), but mountains, sea, broad horizons...

Printed in Great Britain
by Amazon